Reinforced Concrete Design I
Design Sessions

A Student Practice Workbook for Structural Engineering

Peter I. Kattan

Petra Books
www.PetraBooks.com

Peter I. Kattan, PhD

Correspondence about this book may be sent to the author at one of the following two email addresses:

pkattan@petrabooks.com

info@petrabooks.com

Reinforced Concrete Design I - Design Sessions
written by Peter I. Kattan.
ISBN: 979-8-8692-0862-0

Preface

These are the handwritten notes for the Design Sessions for Concrete Design I course that was taught at Applied Science University by Dr. Peter Kattan in the period 2000-2002. The sessions are based on the book "Reinforced Concrete Design " by Wang and Salmon, Third Edition. This book is currently out of print. Students find these sessions useful and it is good to find them in one single volume. The author hopes to make these sessions available to students worldwide and also to revive the Wang and Salmon book. These sessions are for the first course on reinforced concrete design. There are a total of five sessions in this volume.

February 2024	Peter I. Kattan

Reinforced Concrete Design I - Design Sessions
A Student Practice Workbook for Structural Engineering

Contents

Design Session I   I 1

Design Session II   II 1
Service Loads

Design Session III   III 1
Flexural Analysis of Beams

Design Session IV   IV 1
Analysis and Design of T-Beams

Design Session V   V 1
Serviceability - Deflections

Reinforced Concrete Design I - Design Sessions
A Student Practice Workbook for Structural Engineering

Units:

* Two systems of units are used in this course:

(1) <u>SI system:</u>

   length: m ≅ mm ; 1 m = 1000 mm.

   Load : N or kN ; 1 kN = 1000 N.

   Stress : MPa .

$$1 \text{ Pa} = 1 \text{ N/m}^2$$
$$1 \text{ MPa} = 1 \times 10^6 \text{ N/m}^2$$
$$= 1000 \text{ kN/m}^2.$$

   Density : kg/m^3 .

(2) <u>U.S. Customary system:</u> length : in. ≅ ft. ; 1 ft = 12 in.

   Load . $\underset{(pound)}{\underline{lb}}$ or $\underline{kip}.(\underline{k})$, 1k = 1000 lb.

   Stress: psi (lb/in^2)

   ksi ; 1 ksi = 1000 psi.

   Density : lb/ft^3 (pcf).

BASIC STRUCTURAL CONCEPTS:

(1) <u>Forces</u> : Force is measured in $\underline{\underline{N}}$ or $\underline{\underline{lb}}$.
    (Newton)   (pound).

(2) <u>Moments</u> : A moment is a force times a distance.

   Moment is measured in $\underline{\underline{N \cdot m}}$ or $\underline{\underline{lb \cdot ft}}$.

(3) <u>Stress and Strain:</u>

   In <u>uniaxial tension</u>,

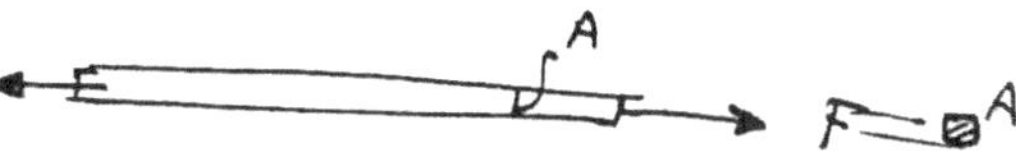

$$\sigma \equiv f \equiv \text{Stress} = \frac{\text{Force}}{\text{Cross-sectional Area}} = \frac{F}{A}$$

   Units of stress are $\underline{\underline{N/m^2}}$ or Pascal (<u>Pa</u>),

     ≅ psi (lb/ft^2) .

   We also have $1 MPa = 1000 \text{ kN/m}^2 = 1 \times 10^6 \text{ N/m}^2$.

               $1 ksi = 1000 \text{ psi}$

Strain $\varepsilon = \dfrac{\text{Extension}}{\text{Original Length}}$

Strain is <u>dimensionless</u> (no units).

* Some people use mm/mm or in/in for strain.

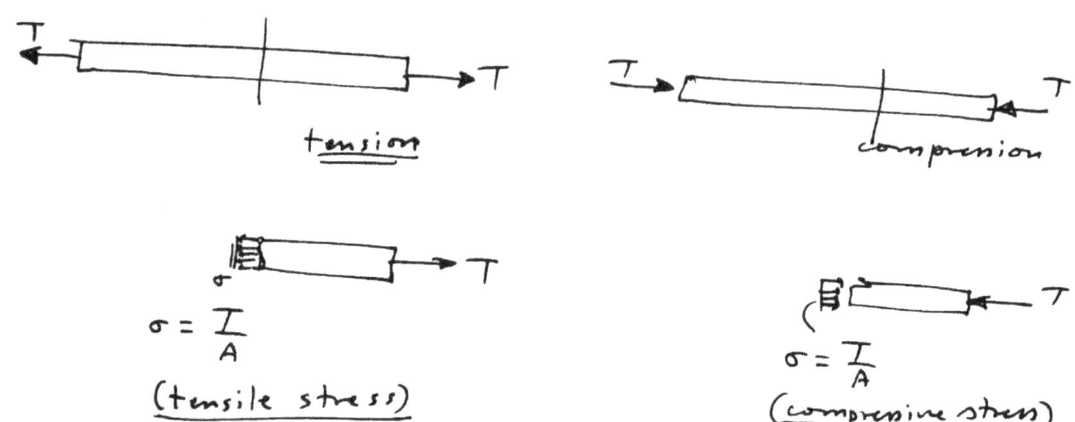

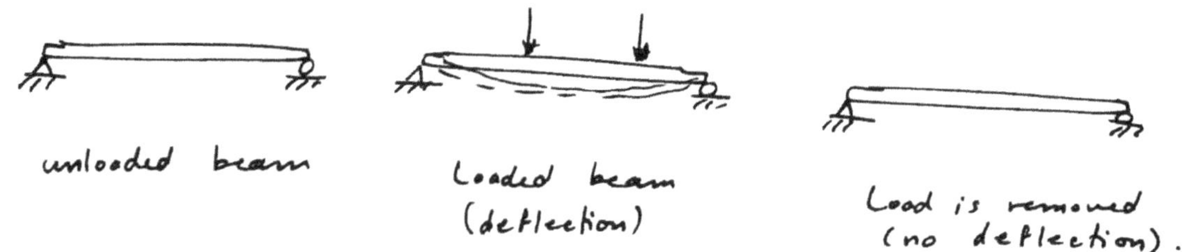

---

(4) Elastic and Plastic Range :

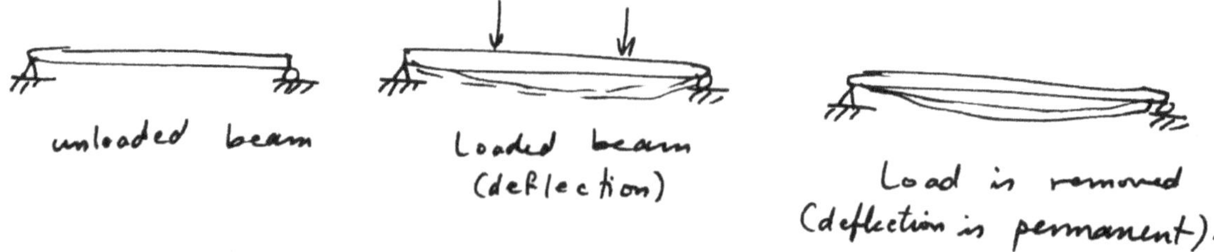

<u>Elastic Behavior</u>: If the beam behaves <u>elastically</u>, it will return to its original position when the load is removed, no matter how many times the load is applied.

<u>Plastic Behavior</u>: Plastic deformation occurs when a beam does <u>not</u> return to its original shape after a load has been applied and removed.

* Every time the load is applied, the plastic deformation will ⌐t3 increase until a point is reached when the beam will <u>fail</u>.

* The point of breaking is called the <u>ultimate strength</u> of the material.

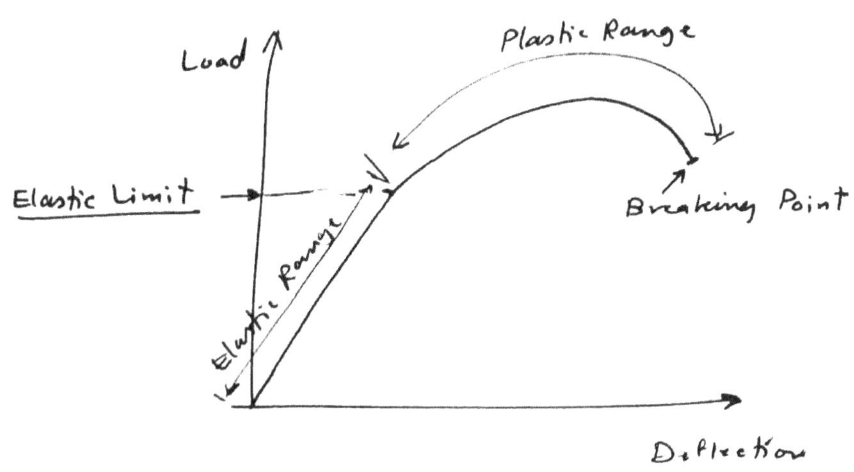

* Elastic materials usually only behave <u>elastically</u> up to a certain load (the <u>elastic limit</u>), after which the material behaves <u>plastically</u>.

* The material will <u>not fail</u> until the breaking point is reached, although the structure may have deformed well before this point, giving ample warning of failure.

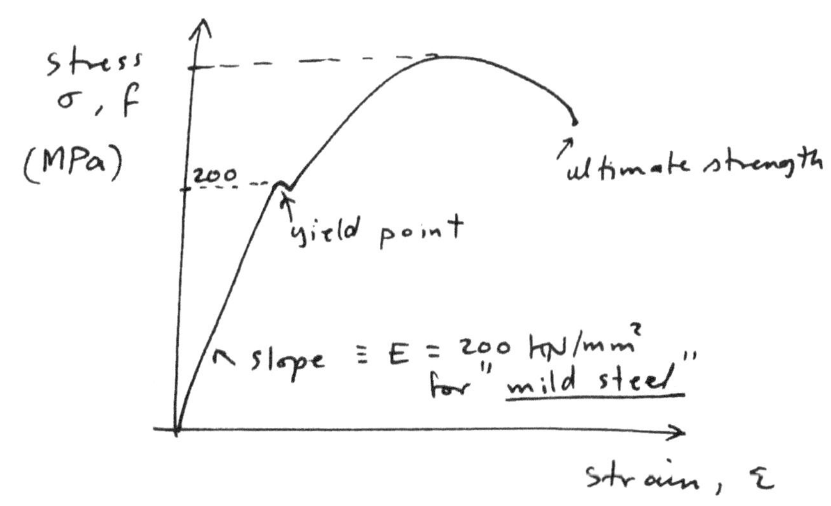

<u>Note:</u>
$$E = 200 \text{ kN/mm}^2 = \frac{200 \times 10^3}{(1 \times 10^{-3})^2} = 200 \times 10^9 \text{ N/m}^2 = 200,000 \text{ MPa}.$$

STRUCTURAL THEORY RELATED TO SIMPLE BEAMS:

① Failure of Beams Due to Bending, Shear, and Deflection:

* The strength will depend upon two forces:
    1. the force which will cause the beam to bend,
    2. the force which will cause the fibers to shear past each other.

* The first force will produce <u>failure due to bending</u>, and the second force will produce <u>failure due to shear</u>.

Failure due to Bending:

* If the tension or compression exceed the strength of the material from which the beam has been made, then the beam will fail due to excessive bending stresses.

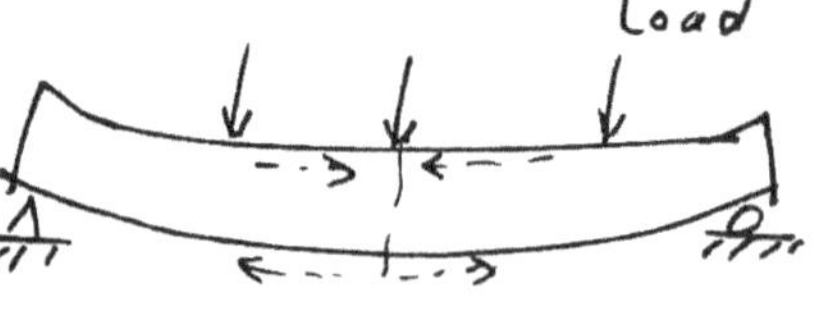

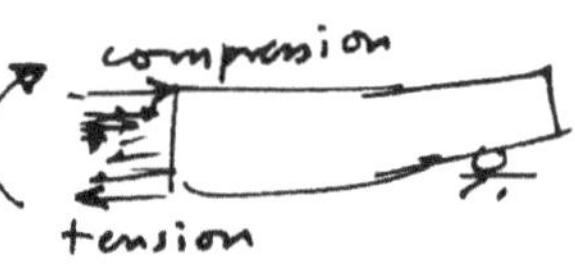

Example: Consider a concrete beam:

* The load on the beam will cause <u>tension cracks</u> in the bottom fibers.

* If the load is increased, the cracks will become larger and the beam will fail.

* This is because concrete has <u>very little strength in tension.</u> (although it is very strong in compression).

Example: the concrete beam must be <u>reinforced</u> with steel reinforcing bars. The steel bars must be placed at the location of the tensile stresses.

* The tensile stresses will be resisted by the steel, and the beam will carry a much greater load before it fails under bending.

* For failure to occur, one of the following two things must happen:

    (1) the concrete at the top is <u>crushed</u>.

    (2) the steel in the bottom is pulled apart (<u>yielded</u>)

* Which one will happen first will depend upon the amount of steel in the bottom of the beam.

    (1) If there is only a small amount of steel, then steel will fail first.

    (2) If there is a large amount of steel, then the concrete in the top will crush long before the steel will begin to yield.

<u>Failure due to Shear:</u>

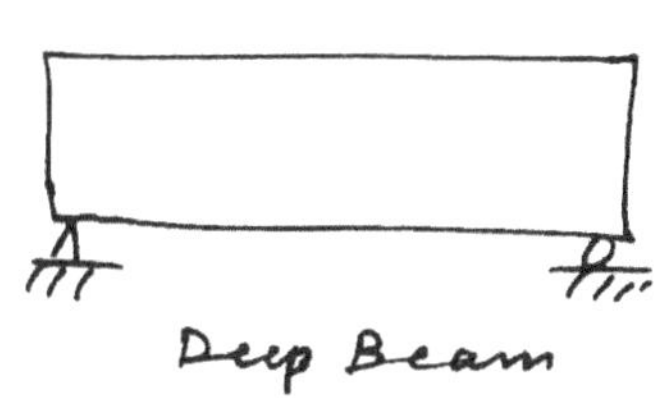

Deep Beam

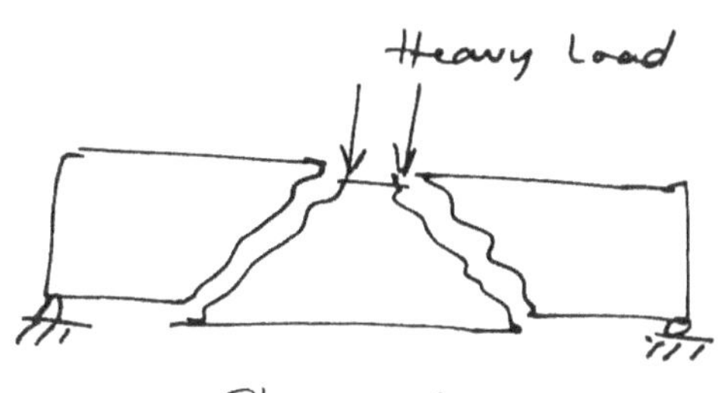

Shear Failure

* Consider a deep beam as shown in the figure.

* If a very heavy load is placed on the beam, it will "punch" a section out of the beam. This kind of failure is called <u>shear failure</u>.

* The angle of the shear line is usually about 45°.

* Steel reinforcement must be placed across the 45°- "shear cracks". This can be done in two ways:

    (1) Use <u>stirrups</u> or <u>links</u> vertically along the beam. These links are vertical reinforcement bars which wrap around the main reinforcement bars in position.

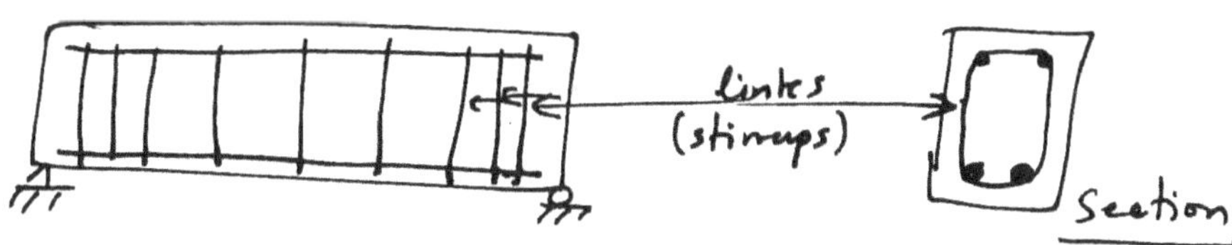

(2) If the shear loads are very large, it may be necessary to increase the shear strength by bending the main reinforcing bars at 45° so that they cross the "shear cracks" at right angles.

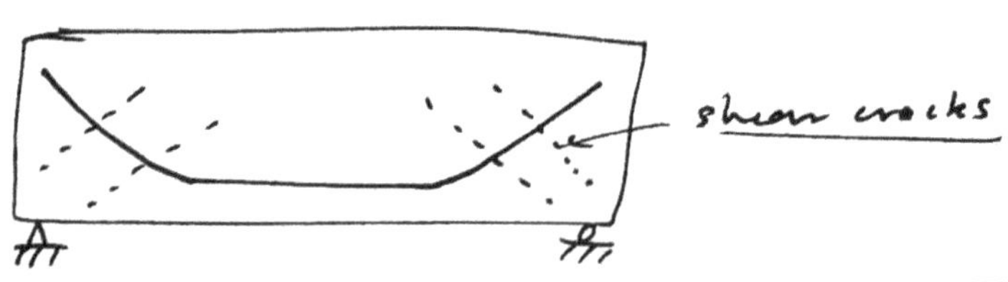

Failure due to Deflection:

* A small load on the beam will cause it to deflect.
* If the load is increased, then the deflection may be <u>excessive</u>, causing the beam to look unsafe, although structurally it is not.
* In addition, excessive deflection may cause damage to the finishings such as the plaster around the beam, and cause the floor to tilt.
* Therefore, from a <u>serviceability</u> or <u>function</u> point of view, the beam has failed due to excessive deflection.

② Bending Moment and Shear Force:

<u>N.A.</u>: Neutral Axis

Bending Moment:

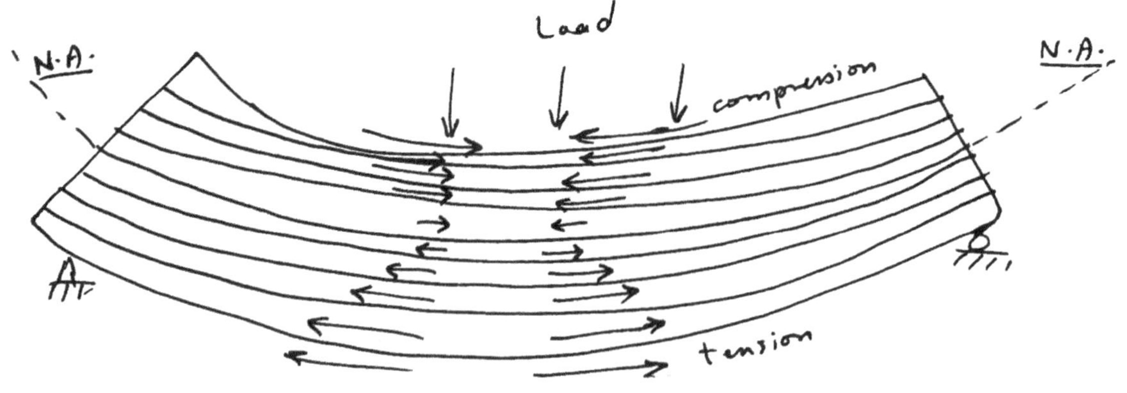

* When a beam bends, the horizontal fibers will change in length. The top fibers will become shorter and the bottom fibers will become longer.

* the load on the beam is resisted by the compression force C in the top and the tension force T in the bottom of the beam.

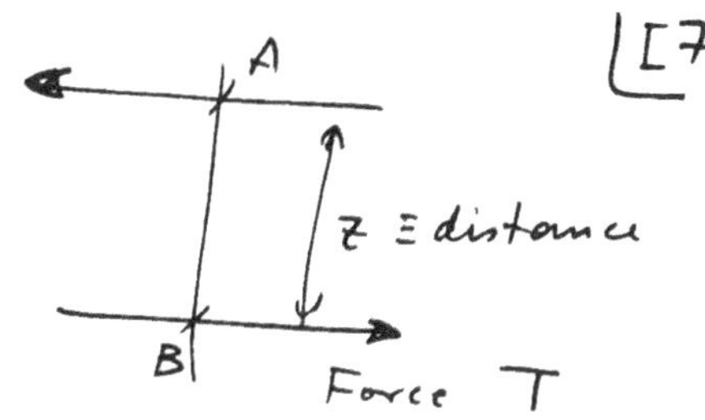

$\xrightarrow{+} \ \Sigma F_x = 0:$ $T - C = 0$ $\Rightarrow$ $T = C$.

$+\circlearrowleft \ M_A = T \cdot z$
$+\circlearrowleft \ M_B = C \cdot z$ } bending moment in the beam.

$\therefore \ M_A = M_B \equiv \underline{\underline{M}}$.

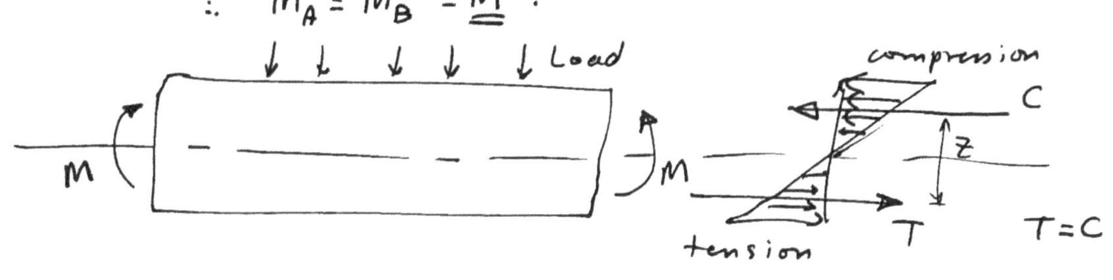

Shear Force :

$V \equiv$ shear force

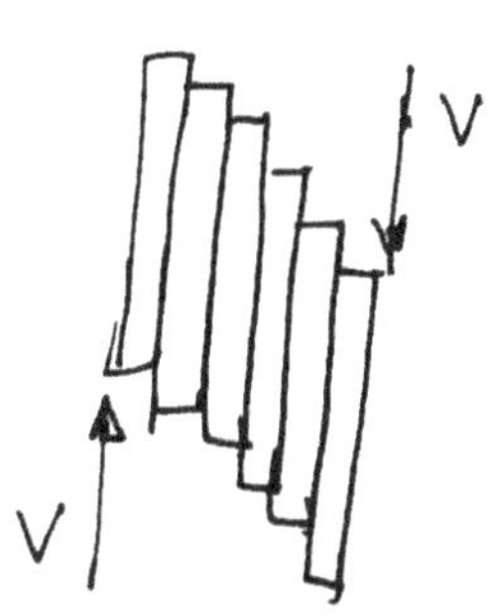

---

(3) <u>Bending-Moment and Shear-Force Diagrams</u>:

<u>Sign Convention</u> :

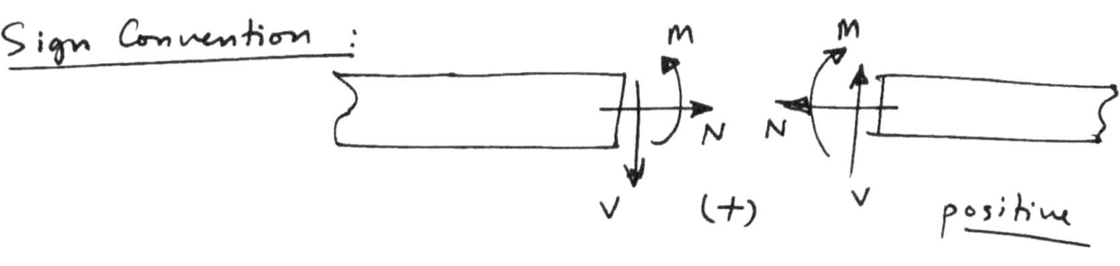

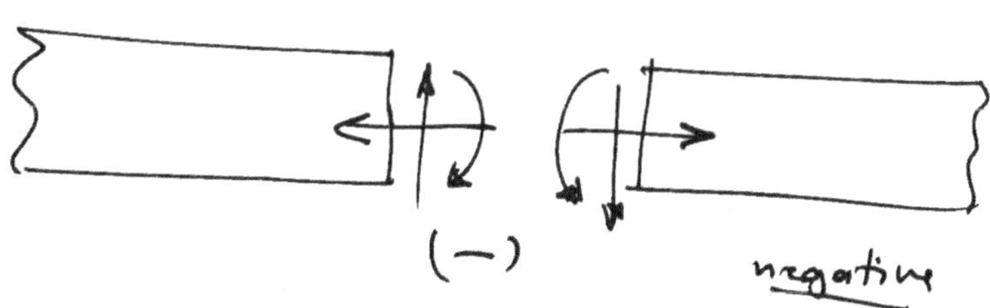

Example:

$$M_{max.} = \frac{PL}{4}.$$

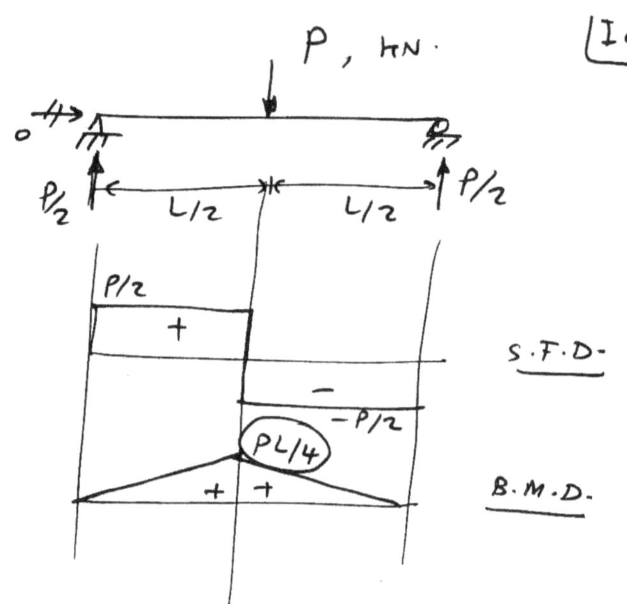

Example:

$$M_{max.} = \frac{1}{2}\left(\frac{wL}{2}\right)\cdot\left(\frac{L}{2}\right)$$

$$\therefore M_{max.} = \frac{wL^2}{8}.$$

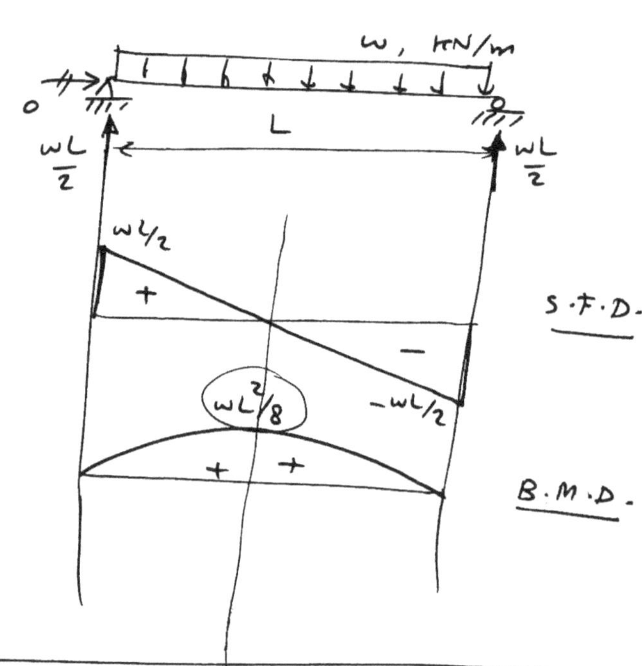

---

④ Laws of Bending:

$f \equiv \sigma \equiv$ stress

$y \equiv$ distance to the neutral axis (N.A.)

$$\boxed{f = \frac{My}{I}}$$

$I \equiv$ second moment of area (moment of inertia)

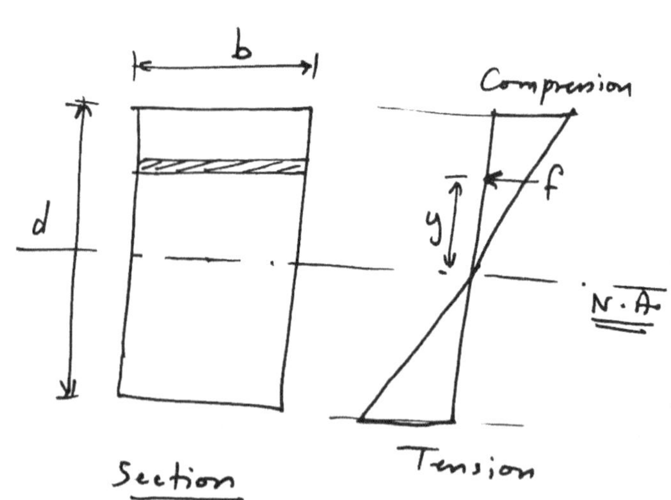

let $Z = \dfrac{I}{y} \equiv$ section modulus.

$$\therefore \quad f = \dfrac{My}{I} = \dfrac{M}{(I/y)} = \dfrac{M}{Z}.$$

Moment of Inertia (Second Moment of Area): I

* Its units are $\underline{\underline{mm^4}}$ or $\underline{\underline{in^4}}$.

Example :

$$\boxed{I_x = \dfrac{1}{12} bd^3}$$

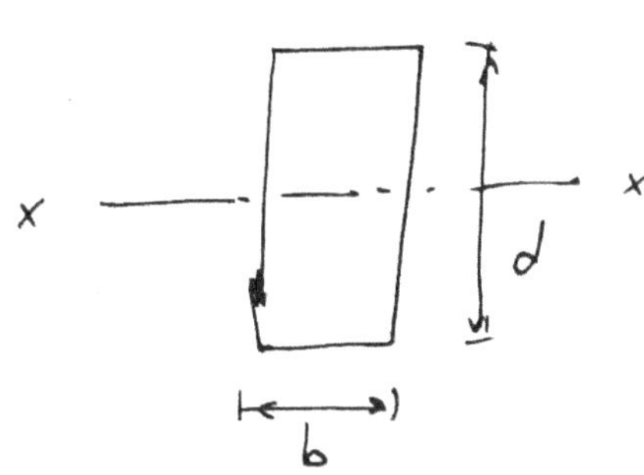

take $b = 4$ mm,
$\quad\quad d = 30$ mm.

$$I_x = \dfrac{1}{12}(4)(30)^3 = 9000 \ mm^4.$$
$$(\underline{very \ high})$$

Example :

take $b = 30$ mm,
$\quad\quad d = 4$ mm.

$$I_x = \dfrac{1}{12} bd^3 = \dfrac{1}{12}(30)(4)^3$$
$$= 160 \ mm^4.$$
$$(\underline{very \ low})$$

Note that $\dfrac{9000}{160} \simeq 56.25 \simeq \underline{\underline{56}}.$

* The <u>ruler</u> is approximately 56 times stiffer in the stiff direction than in the flexible direction.

---

⑤ <u>Calculating Beam Deflections:</u>

* It is very important to limit deflections in building structures so that the function or serviceability of the building is not impeded.

* The deflection depends on the following factors: [I 10

  1. Load , w .   (kN/m) · [Uniformly Distributed Load].
  2. Span , L .   (m) .
  3. Modulus of Elasticity , E (kN/m^2) .
  4. Moment of Inertia , I (m^4) .

* In general ,
$$\text{Deflection} = \text{constant} * \frac{wL^4}{EI}$$

* If we have a concentrated load P, then :
$$\text{Deflection} = \text{constant} * \frac{PL^3}{EI}$$

Example :
$$\delta_{max.} = \frac{1}{48}\frac{PL^3}{EI}$$

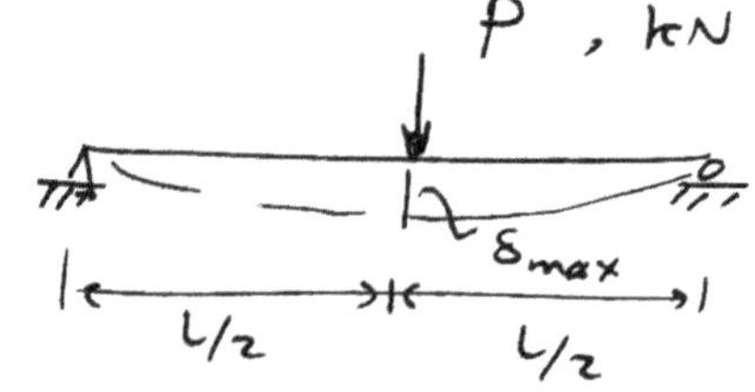

Example :
$$\delta_{max.} = \frac{5}{384}\frac{wL^4}{EI}$$

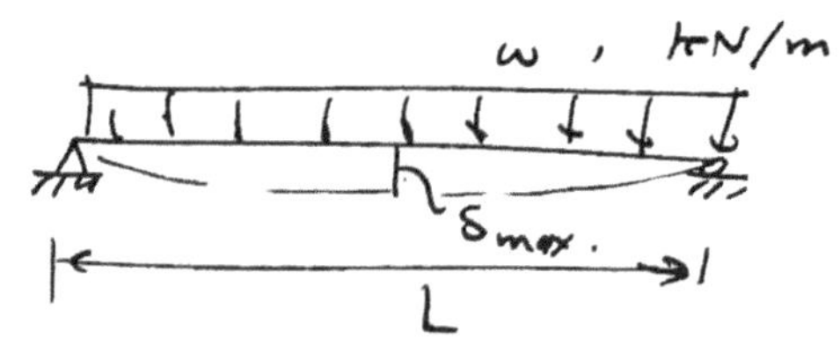

Problem 1 :
Draw the shear force and bending moment diagrams
for the overhanging beam shown in the figure.

Solution :
$$A_y = B_y = \frac{2+2+6}{2} = 5 \text{ kN} \uparrow .$$

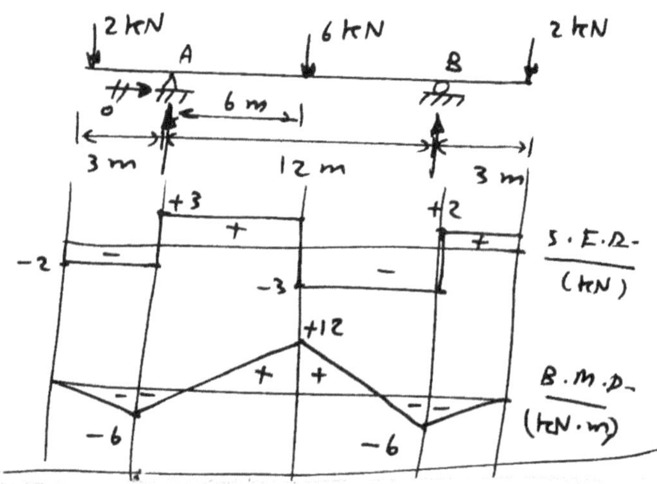

Problem 2 :

Draw the shear force and bending moment diagrams for the overhanging beam shown in the figure.

Solution :

$$A_y = B_y = \frac{(1)(18)}{2} = 9 \; kN \uparrow \cdot$$

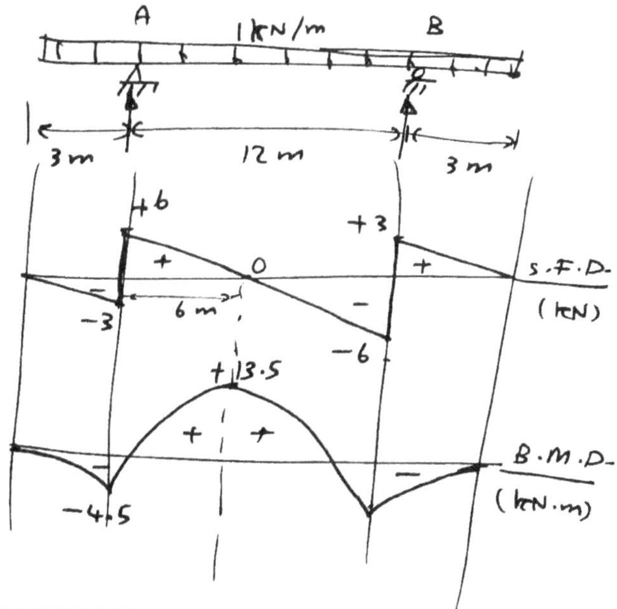

Problem 3 :

A ruler 300 mm long and with a cross-section of 30 mm × 4 mm is placed between two tables 280 mm apart. If a 10 N load is placed in the middle, what are the maximum stresses in the ruler when:

(a) the ruler is laid flat.
(b) the ruler is held on its side.

Solution :

$$\text{stress} \quad \sigma = f = \frac{My}{I}$$

$$M_{max.} = \frac{PL}{4} = \frac{(10)(280)}{4}$$

$$= 700 \; N \cdot mm$$

(a) the ruler is laid flat :

$$I_x = \frac{1}{12}(30)(4)^3 = 160 \; mm^4 \cdot$$

take $y = 2 \; mm$

stress $f = \dfrac{My}{I} = \dfrac{(700)(2)}{160} = \underline{\underline{8.75}} \ N/mm^2$.

(b) The ruler is held on its side :

$I_x = \dfrac{1}{12}(4)(30)^3 = 9000 \ mm^4$.

   take $y = 15 \ mm$

stress $f = \dfrac{My}{I} = \dfrac{(700)(15)}{9000} = \underline{\underline{1.167}} \ N/mm^2$.

Problem 4 :

Find the deflection of the ruler in the previous problem when the ruler is laid flat. Take $E = 8 \ kN/mm^2$.

Solution :

Deflection $\delta_{max.} = \dfrac{1}{48}\dfrac{PL^3}{EI} = \dfrac{1}{48}\dfrac{(10)(280)^3}{(8000)(160)} = 3.57 \ mm$

$$\approx \underline{\underline{3.6 \ mm}}.$$

Problem 5 :

Find the deflection of the ruler in the previous problem when the ruler is held on its side.

Solution :

Deflection $\delta_{max.} = \dfrac{1}{48}\dfrac{PL^3}{EI} = \dfrac{1}{48}\dfrac{(10)(280)^3}{(8000)(9000)} = 0.0635 \ mm$

$$\approx \underline{\underline{0.06 \ mm}}.$$

Common U.S. Customary Units, Their SI Equivalents, and SI Prefixes Used in Static Structural Analysis

U.S. Customary Units and SI Equivalents

Measure	U.S. Customary Unit	SI Equivalent
Length	in	25.4 mm
	ft	0.3048 m
Area	in^2	$645.2 \ mm^2$
	ft^2	$0.0929 \ m^2$
Volume	in^3	$16.39 \ cm^3$
	ft^3	$0.02832 \ m^3$
Force	lb	4.448 N
	kip	4.448 kN
Stress	lb/in^2	$6.895 \ N/m^2 = 6.895 \ Pa$
	lb/ft^2	$47.88 \ N/m^2 = 47.88 \ Pa$
Moment of force	$in \cdot lb$	$0.1130 \ N \cdot m$
	$kip \cdot ft$	$1.356 \ kN \cdot m$
Moment of inertia (of an area)	in^4	$41.62 \times 10^4 \ mm^4$

Common SI Prefixes

Factor	Prefix	Symbol
10^9	giga	G
10^6	mega	M
10^3	kilo	k
10^{-3}	milli	m
10^{-6}	micro	μ
10^{-9}	nano	n

lb/ft^3 $157.080 \ N/m^3$

kip/ft $14.593 \ kN/m$

Service Loads

Example 1:

The floor beam shown in the figure is used to support the 2-m width of a reinforced lightweight concrete slab having a thickness of 100 mm. The slab serves as a portion of the ceiling for the floor below and therefore its bottom is coated with plaster. Furthermore, a 2.5-m high, 300 mm light aggregate concrete block wall bears directly on the beam. Determine the loading on the beam measured per meter of length of the beam.

Solution: Get all the materials from the Dead Load Tables)

(1) Concrete Slab (Reinforced Lightweight Concrete):

(25)(20) kN/m^3 (from tables)

$$\therefore \quad 20 \times 2 \times \frac{100}{1000} = \boxed{4 \ kN/m} \cdot \quad 5 \ kN/m$$

(2) Plaster Ceiling: $500 \ N/m^2 \equiv 0.5 \ kN/m^2$.

$$0.5 \times 2 = \boxed{1 \ kN/m}$$

(3) Block Wall (light aggregate concrete):

plain concrete $\rightarrow$ 23 kN/m^3.

$$23 \times 2.5 \times \frac{300}{1000} = \boxed{17.25 \ kN/m}$$

$$\therefore \text{Total Load on Beam} = (4) + 1 + 17.25 = (23).25 \ kN/m.$$

$$D = 22.25 \ kN/m$$

Live Load Reduction:

* For some types of structures, many codes will allow a _reduction_ in the uniform live load for a <u>floor</u> since it is unlikely that the prescribed live load will occur simultaneously throughout the entire structure at any one time.

* For example, ANSI A58.1-1982 allows a reduction of live load on a member having an <u>influence area</u> of $400 ft^2$ ($\approx 37 m^2$) or more. This reduced live load is calculated using the following equation:

$$L = L_o \left(0.25 + \frac{15}{\sqrt{A_I}} \right)$$

where

L : reduced design live load per square foot of area supported by the member.

L_o : unreduced design live load per square foot of area supported by the member.

A_I : influence area in ft^2, equal to four times the tributary or effective load-carrying floor area for a column, and two times the tributary or effective load-carrying floor area for a beam.

* The reduced live load defined by the above equation is limited to not less than 50% of L_o for members supporting one floor, or not less than 40% of L_o for members supporting more than one floor. <u>No</u> reduction is allowed for structures used for public assembly, garages, or roofs.

Example 2 :

A two-story office building has interior columns that are spaced 6.7 m apart in two perpendicular directions. If the (flat) roof loading is $1 \, kN/m^2$, determine the reduced live load supported by a typical interior column located at ground level.

Solution:

* A_T: Tributary Area of an interior column

$$A_T = (6.7)(6.7) = 44.89 \ m^2.$$

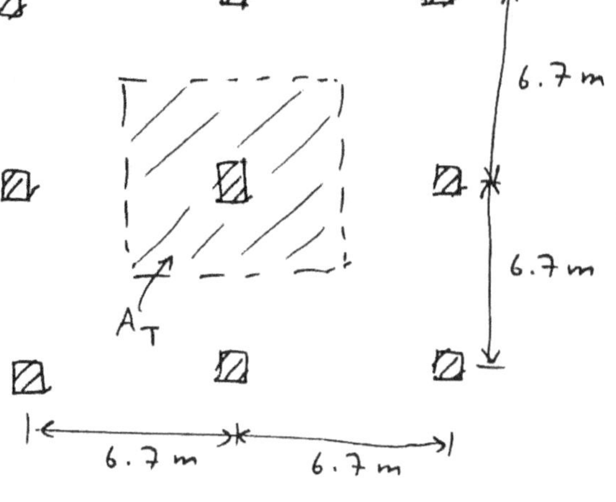

The roof load on the ground-floor column

is $(1)(44.89) = \underline{44.89 \ kN}.$

This load is <u>not</u> reduced because it is <u>not</u> a <u>floor</u> load (it is a <u>roof</u> load).

* For the second floor, the live load is obtained from the tables. Take $L_o = 2.5 \ kN/m^2.$

Check if we can reduce this live load.

For columns, influence area $A_I = 4 A_T = 4(44.89)$
$$= 179.56 \ m^2$$
$$> 37 \ m^2. \ \checkmark$$

∴ We can reduce the live load L_o.

* Convert units to use the formula:

$$A_I = 179.56 \ m^2 = 1933 \ ft^2.$$

$$L_o = 2.5 \ kN/m^2 = \frac{2.5 \times 1000}{47.88} = 52.2 \ \ lb/ft^2.$$

$$\therefore L = L_o \left(0.25 + \frac{15}{\sqrt{A_I}}\right) = 52.2 \left(0.25 + \frac{15}{\sqrt{1933}}\right) = 30.86 \ lb/ft^2$$
$$= \frac{30.86 \times 47.88}{1000}$$
$$= 1.48 \ kN/m^2.$$

% Load Reduction $= \frac{1.48}{2.5} \times 100\% = 59.2\% > 50\%. \ \underline{o.k.}$

∴ the floor load on the ground-floor column is

$$(1.48)(44.89) = \underline{66.43} \ kN.$$

* Total Live Load Supported by the ground-floor column is:
$$44.89 + 66.43 = \underline{111.32} \ kN. \ (reduced \ live \ load).$$

Example 3 :

Consider the gabled frame shown in the figure. Assume that the structure is located at 1000 m above sea level. Determine the roof snow load acting on a horizontal projection.

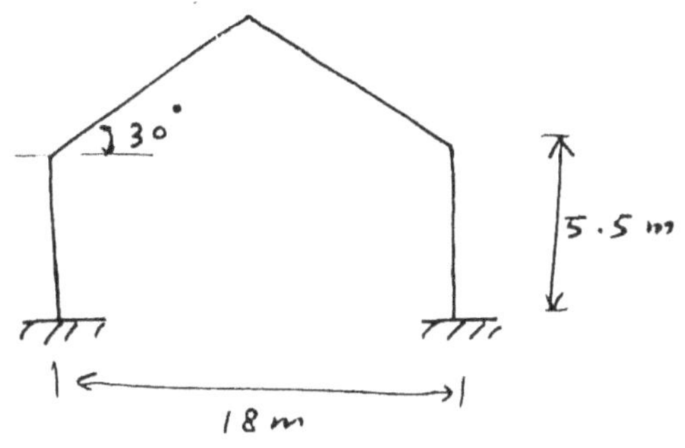

Solution :

Elevation above sea level $h = 1000$ m.

$$\text{Snow load} = \frac{h - 400}{400} = \frac{1000 - 400}{400} = 1.5 \text{ kN/m}^2.$$

$\alpha = 30° > 25°$. We need to reduce the snow load.

$\therefore$ Reduced snow load $= (0.90)(1.5) = \underline{1.35 \text{ kN/m}^2}$

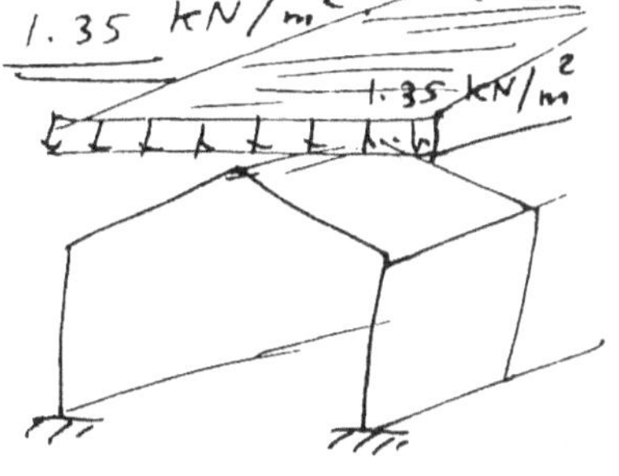

Example 4 :

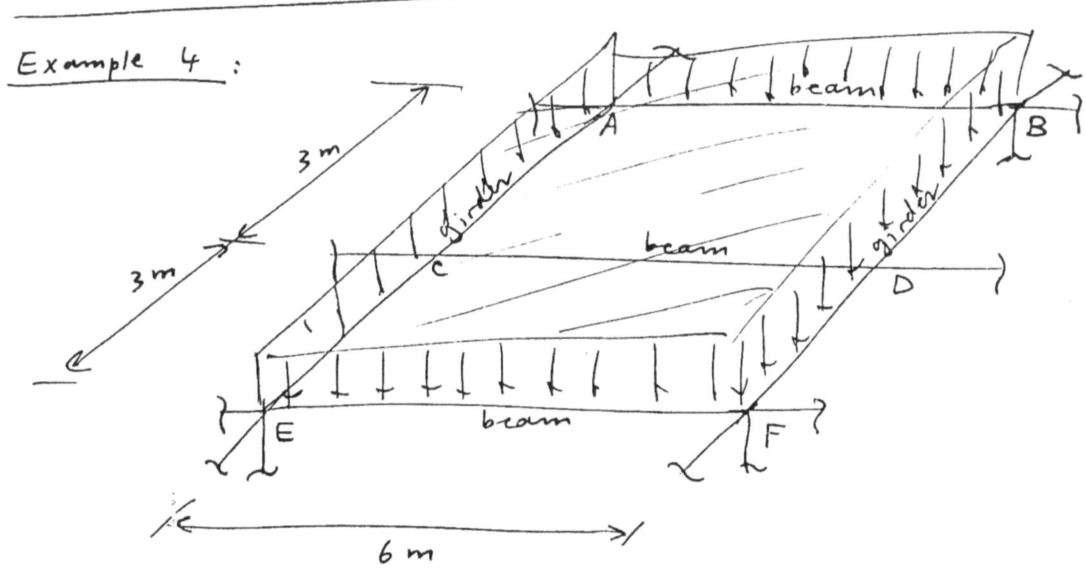

For the floor system shown in the figure, assume that a surface load 4.8 kN/m² is applied on panel ABEF.

Determine the resulting effects on beams AB, CD, [#5]
EF, girder AE and the columns at A and E.

Solution:

(1) Beams AB and EF:

Tributary Area = 1.5 × 6
= 9 m².

Uniform Load ω = 4.8 × 1.5
= 7.2 kN/m.

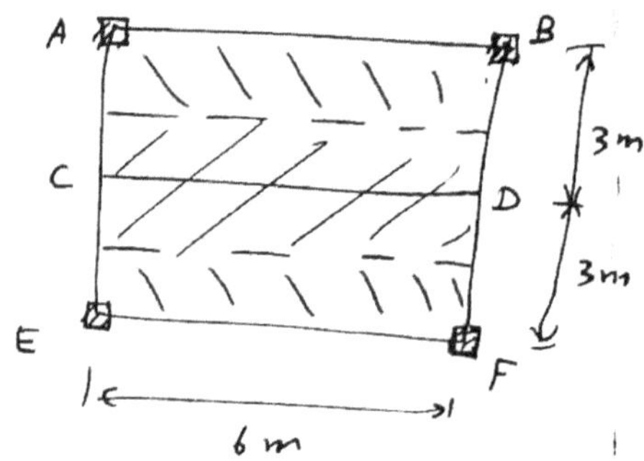

By symmetry,
End Support Load = $\dfrac{\omega L}{2}$ = $\dfrac{(7.2)(6)}{2}$
= 21.6 kN.

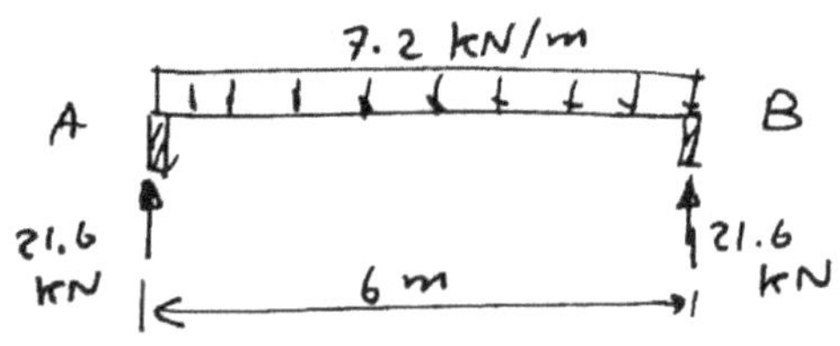

(2) Beam CD:

Tributary Area = 3 × 6 = 18 m².

Uniform Load ω = 4.8 × 3 = 14.4 kN/m.

By symmetry,
End Support Load = $\dfrac{\omega L}{2}$ = $\dfrac{(14.4)(6)}{2}$
= 43.2 kN.

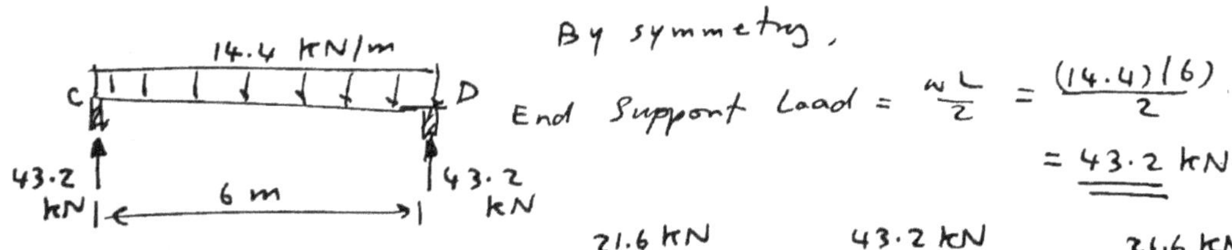

(3) Girder AE:

By symmetry,

End Support Load = $\dfrac{\text{Total Load}}{2}$

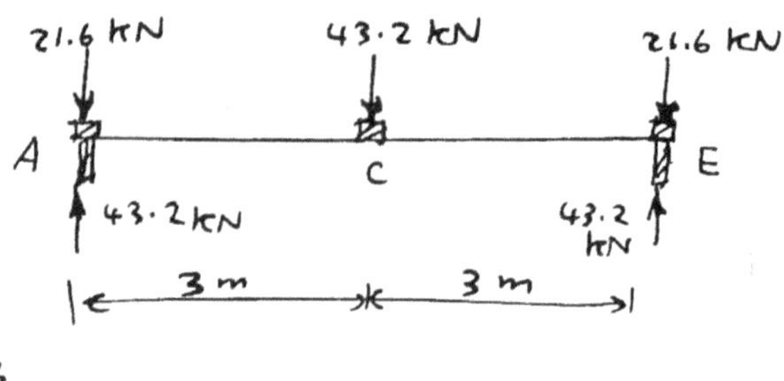

$= \dfrac{21.6 + 43.2 + 21.6}{2}$ = 43.2 kN.

(4) Column Loads at A and E:

* The girder end supports loads for girder AE are applied to the column tops.

∴ Column load at A and E = 43.2 kN.

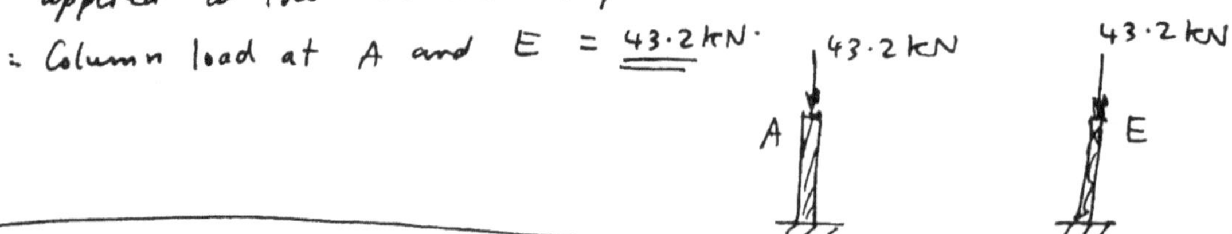

LOADS

DEAD LOADS

BUILDING MATERIALS

MATERIAL		kN/m^3
Aluminium		28
Asphalt		13
Bricks	mud - broken	10 - 14
	Burnt	16 - 19
Cast - iron		72.5
Cement		16.0
Glass		25 - 27
Gravel		15 - 18
Lime (slaked)		5.5 - 14.9
Stones		
	Basalt	30
	Granite	28
	Marble	27
	Lime	15.0
Fine aggr. (Sand)		13.9
Coarse (Broken) aggregate:		14.5
Timber (Wood)	Hard	6 - 7.2
	Soft	3.9 - 5.3
SOILS		
Clay - Loose - dry		14
	compact - dry	17
	compact - wet	20.5
Earth - Loose - dry .		12.5
	compact - dry	15.0
	Loose - moist	15 - 16
	compact - moist	18 - 19
GRAVEL - Loose		16
	- Rammed	19 - 21
SAND - clean - dry		15 - 16
	River	18
	Wet	17 - 20
SILT - Fine - dry		16
	Fine saturated	20.5

MATERIAL		N/m^2
Galvanised iron 1mm thick sheet		85
Tiles	ceramic (20-25 mm)	600
	Country	600

CONCRETE

Type	kN/m^3
Light weight concrete	12 - 20
Lime concrete	19
Plain Concrete	23
Reinforced concrete (with 2% steel)	25

FLOORING

Type	N/m^2
Cement-concrete - Standard	600
ceramic tiles (20 mm)	500
Mortar screeding (10 mm)	200
Terrazo (10 mm)	200
Tiled flooring on 25 mm thick cement mortar	1000
Tiled flooring on 25 mm thick cement mortar over 50 mm thick sand layer	2000

MASONRY

Type		kN/m^3
Ashlar -	Granite	26
	Lime stone	25
Rubble -	Dry stone	20
	Granite	22
Stone masonry		
	Basalt	30
	Granite	28
	Marble	27
	Sand	23

MORTARS

Type	kN/m^3
cement mortar (1:3 to 1:6)	21 - 23
Lime mortar (1:1 to 1:3)	18 - 20
Mud mortar	15 - 18

DEAD LOADS (contd.)

PLASTERING

Type	N/m²
cement Plaster	
25 mm	500
10 mm	200
ceiling plaster	
10 mm	250
20 mm	500
Mortar screeding 10mm	200

ROOFING

Type	N/m²
Asbestos sheeting	
Corrugated	160
plain	120 - 150
Galvanised iron sheets (18 - 22 gauge)	150
Glazing (6 mm)	300
Mortar Screeding	200
Bituminous felt (10 mm)	100
Brick-bat coba (75mm)	1200-1400
Slates (10 mm)	300
Stone slabs for each 10 mm thick	300

WALLS

Type	kN/m³
Brick masonry walls	
Mud brick	18 - 20
Hollow cement block	14 - 19
Concrete Walls	
plain concrete	23
reinforced	25
Stone masonry walls	
Basalt	30
Lime stone	27
Sand stone	23

PARTITION WALLS 100mm thick

Additional load on floors due to 100 mm thick partition Walls	1000 N/m²

SOILS

Type	angle of repose	unit. wt. kN/m³
Clay		
dry	29°	17
damp	45°	18
saturated	15°	19
Gravel		
clean	48°	17.5
sandy	26°	19
Sand		
Loose - dry	31° - 37°	16
Compact - dry	35°	19
Wet	32°	18
Very wet	26°	19

Type	co-efficient of friction on concrete μ
Clay, gravel Sand - dry	0.5 — 0.6
Clay - Sand - moist	0.33
Clay - Wet	0.1 — 0.3
Rock	
- dry	0.6 - 0.7
- Wet	0.5

Type	Safe bearing capacity kN/m²
Rocks	1500 and over
Sand, gravel - firm	600 - 800
Clay - hard	400 - 600
Clay - Sand - dry	300 - 400
Clay - ordinary - dry	200 - 300
Clay - soft	100 - 200

LIVE LOADS ON
ROOFS & FLOORS

Type of building		Uses of floor	u.d.l	Concen trated
General	Particular		KN/m^2	KN.
RESIDENDIAL PREMISES	Flats Houses Bunglows upto one storey	All rooms including bedrooms Kitchens Laundries	2.0	1.4
	Hotels motels Hospitals	Bedrooms including hospital wards	2.0	1.8
	Boarding houses, Hostels	Bed rooms & Halls including dormitories	2.0	1.8
PUBLIC BUILDINGS	Public halls Theatres cinemas Assembly halls in schools Collages Sports halls	with fixed seating	4.0	–
		without fixed seating	5.0	3.6
	Dance halls Drill halls gymnasia	–	5.0	3.6
	Mosques Churches	Places of worship chapels	3.0	2.7
	class rooms	not exceeding 40 sq.m. in area	3.0	2.7
	Library reading rooms	with book storage	4.0	4.5
		without book storage	2.5	4.5

Type of building		Uses of floor	u.d.l	Concen trated
General	Particular		KN/m^2	KN.
PUBLIC BUILDINGS	Library	Book stores	6.0	–
	Museums & art galleries	–	4.4	–
	Hotels Banking halls	bars –	5.0 4.0	– –
	Stores shops Departmental stores	Display & sale	4.0	3.6
	Miscallaneous	Dressing rooms in gymnasia theatres colleges	2.0	1.2
		Baths, water closets. Toilets	2.0	1.8
COMMERCIAL AND INDUSTRIAL PREMISES	offices	General	2.5	2.7
		Filing & storage	5.0	–
		Computer rooms	3.5	–
	Theatres Cinemas studios (T.V. Radia stations)	studios	4.0	–
		Stages in theatres	7.5	–
		Stages-gymnasia & colleges	5.0	–
		Cantilever Balconies	4.5	–
		Projection rooms (incl. Equipment	5.0	–

LIVE LOADS (contd)

Type of building		Uses of floor	udl	Conc.
General	Particular		kN/m^2	kN
COMMERCIAL AND INDUSTRIAL PREMISES.	Work places Factories etc:	X-ray utility operating rooms & Clinics	2.0	4.5
		Laundries	3.0	4.5
		Kitchens incl. Normal equipment	$\not< 3.0$	4.5
		Laboratories (incl. Eqpt.)	$\not< 3.0$	$\not< 4.5$
		Light work rooms without storage	2.5	1.8
		workshops Light / Medium / Heavy	5.0 / 7.5 / 10	—
		Foundries & Machine halls	$\not< 20$	—
		Printing places (press)	$\not< 12.5$	—
		paper storage Book stores	$\not< 12.5$	—
		storage of material for printing	4	—
	miscellaneous	Motor rooms fan rooms (incl. Eqpt)	$\not< 7.5$	—
		Boiler rooms	7.5	—
	Corridors passages Hallways	subject to Crowds	4.0	45
		over crowding	$\not< 5.0$	$\not< 4.5$

Stairs & Landings	Dwellings → 3 storeys / other buildings / Landings	2.0 / 3.0 / 5.0	or LL as on floors.

HORIZONTAL LIVE LOADS ON PARAPETS & BALUSTRADES.*

Description		kN/m
Light access stairs	$\not> 600mm$	0.50
gangways	"	0.75
other stairs Landings Balconies	Domestic & private	1.00
	others	
parapets & guard rails	on roofs	1.00
Panic barriers		3.00

* Loads act horizontally at level of handrail or Coping

SNOW LOADS*

Elevation of structure above sea Level	kN/m^2
$h < 250 m$	0
$h > 250 < 500$	$\dfrac{(h-250)}{1000}$
$h > 500 < 1500$	$\dfrac{(h-400)}{400}$
$h > 1500 < 2500$	$\dfrac{(h-812.5)}{250}$

* Snow Loads where applicable on roofs upto slopes of $50°$
For each one millimetre thick snow $= 3 N/m^2$

Description	kN/m^2
Garages, Light Vehicles	4
Medium "	6
Heavy "	7.5

WIND LOADS

Basic windspeed considered for obtaining wind pressures on obstructed areas. is

$$V = 35\,m/sec \quad (126\,Km/hr)$$

(This is measured at 10m above G.L for not less than 3 seconds)

V_z = Design wind speed
$$= V \times S_1 \times S_2 \times S_3$$

where

S_1 = Topography Factor

S_2 = Ground roughness factor

S_3 = coefficient of statistics. related to life of structure

CHARACTERISTIC wind pressure

$$q = 0.613\,V_z^2$$

P = the intensity of wind pressure on any area obstructing wind load

$$= \left(C_e - C_i\right). V \quad N/m^2$$

where

C_e = External pressure coefficiant

C_i = internal pressure coefficiant.

TOPOGRAPHY FACTOR S_1

Description	S_1 .
sites adversely affected by very exposed hill slopes and crests (where wind acceleration is known)	1.1

Continuously exposed sites - (sea shores) to storms.	1.0
Protected sites to Storms (steep-sided Valleys)	0.9

GROUND ROUGHNESS FACTOR. = S_2

This factor depends on type of ground on which the structure is situated type of structure, height of structure exposed to wind forces above ground and wind breaking.

CLASSIFICATION OF GROUND

GROUP	Description
A	open areas - obstructions $< 1.5m$, level areas and deserts
B	open areas - obstructions $< 10m$, air ports villages . suburbs
C	areas with closely spaced obstructions cities , towns
D	Areas with large and frequent obstructions - Capital cities .

WIND LOADS
(contd)

CLASSIFICATION OF TYPE OF STRUCTURE.

Type	Description
I	Single, isolated
II	structures of maximum height or width < 50m
III	structures max. height or width > 50m

COEFFICIENT OF STATISTICS - S_3

Type of structure	S_3
Important buildings, Houses, Hospitals - Power houses, emergency buildings, dams	1.05
Less important buildings, towers in forest areas - Farms	0.95
Temporary Construction	0.83
All other structures	1.00

INTERNAL PRESSURE COEFFICIENT, C_i

Description	All faces impermeable (air tight)	Two opposite faces permeable & other two impermeable
Impermeable **face** & perpendicular to wind.	−0.30	−0.30
Permeable face & Perpendicular to wind	−0.20	−0.20

WIND LOADS (contd)

GROUND ROUGHNESS FACTOR - S_2

Classification of Ground	Group A			Group B			Group C			Group D		
Type h Height	I	II	III	I	II	III	I	II	III	I	II	III
upto 3m	0.83	0.78	0.73	0.72	0.67	0.63	0.64	0.60	0.55	0.56	0.52	0.47
5 m	0.88	0.83	0.78	0.79	0.74	0.70	0.70	0.65	0.60	0.60	0.55	0.50
10	1.00	0.95	0.90	0.93	0.88	0.83	0.78	0.74	0.69	0.67	0.62	0.58
15	1.03	0.99	0.94	1.00	0.95	0.91	0.88	0.83	0.78	0.74	0.69	0.64
20	1.06	1.01	0.96	1.03	0.98	0.94	0.95	0.90	0.85	0.79	0.75	0.70
30	1.09	1.05	1.00	1.07	1.03	0.98	1.01	0.97	0.92	0.90	0.85	0.79
40	1.12	1.08	1.03	1.10	1.06	1.01	1.05	1.01	0.96	0.97	0.93	0.89
50	1.14	1.10	1.06	1.12	1.08	1.04	1.08	1.04	1.00	1.02	0.98	0.94
60	1.15	1.12	1.08	1.14	1.10	1.06	1.10	1.06	1.02	1.05	1.02	0.98
80	1.18	1.15	1.11	1.17	1.13	1.09	1.13	1.10	1.06	1.10	1.07	1.03
100	1.20	1.17	1.13	1.19	1.16	1.12	1.16	1.12	1.09	1.13	1.10	1.07
120	1.22	1.19	1.15	1.21	1.18	1.14	1.18	1.15	1.11	1.15	1.13	1.10
140	1.24	1.20	1.17	1.22	1.19	1.16	1.12	1.17	1.13	1.17	1.15	1.12
160	1.25	1.22	1.19	1.24	1.21	1.18	1.21	1.18	1.15	1.19	1.17	1.14
180	1.26	1.23	1.20	1.25	1.22	1.19	1.23	1.20	1.17	1.20	1.19	1.16
200	1.27	1.24	1.21	1.26	1.24	1.21	1.24	1.21	1.18	1.22	1.21	1.18

WIND LOADS (contd)

WIND LOADS (contd)

EXTERNAL PRESSURE CoEFFICIENT C_e								
Ratio h/w	Ratio L/w	PLAN	ELEVATION	α	C_e			
					a	b	c	d
$h/w \le 0.5$	$L/w > 1 \le 1.5$			$0°$	+0.7	−0.2	−0.5	−0.5
				$90°$	−0.5	−0.5	+0.7	−0.2
	$L/w > 1.5 \le 4$			$0°$	+0.7	−0.25	−0.6	−0.6
				$90°$	−0.5	−0.5	+0.7	−0.1
$h/w > 0.5 \le 1.5$	$L/w > 1 \le 1.5$			$0°$	+0.7	−0.25	−0.6	−0.6
				$90°$	−0.6	−0.6	+0.7	−0.25
	$L/w > 1.5 \le 4$			$0°$	+0.7	−0.3	−0.7	−0.7
				$90°$	−0.5	−0.5	+0.7	−0.1
$h/w > 1.5 \le 6$	$L/w > 1 \le 1.5$			$0°$	+0.8	−0.25	−0.8	−0.8
				$90°$	−0.8	−0.8	+0.8	−0.25
	$L/w > 1.5 \le 4$			$0°$	+0.7	−0.4	−0.7	−0.7.
				$90°$	−0.5	−0.5	+0.8	−0.1

LOADS DUE TO IMPACT & VIBRATIONS

In the case of structural members, carrying Live loads which induce impact or vibration, the increase in Live loads shall be considered in the design.

For concentrated Live Loads with impact or vibration, a minimum impact factor not less than 20% shall be considered

Type of structure	Impact Factor in %
(i) For frames supporting lifts and hoists	100
(ii) For foundations, footing piers supporting lifts and hoisting apparatus	40
(iii) For light machinery shaft or motor units	minimum 20
(iv) For reciprocating machinery or Power units	minimum 50

SEISMIC OR EARTHQUAKE LOADS

Nature of Earthquakes

Earthquakes consist of various forms of waves orginating at the centre of disturbance and causing horizontal and vertical ground movements or vibrations. The movements are complicated due to forced and superimposed vibrations. For simplification of design these movements are divided into vertical and horizontal vibrations. The horizontal vibrations are generally much greater than the vertical ones hence it is these vibrations which are mainly considered in designing earthquake resisting structures.

While a severe earthquake always leaves in its wake many collapsed buildings and much structural damage, it also shows that a properly designed structure is capable of withstanding the strongest recorded earthquake shocks. Furthermore, in concrete this can be accomplished at very little, if any, added expense. This is particularly true in small buildings.

MEASUREMENT OF SEISMIC FORCES

Usually two types of scales are used to measure the intensity of Earthquake effect. (i) Rossi-Forel scale (ii) Richter's Scale.

For the relative measurement of the destructive effort of an earthquake, the Rossi-Forel scale is normally used. This scale which is described in *Table* is divided into 10 points. The first 6 points on the scale do not cause damage to structures.

Provision in the design of structures for the effects of earthquake is made by applying a horizontal acceleration expressed in terms of g, the acceleration due to gravity. The relation between the Rossi-Forel scale and the horizontal acceleration is given in *Table*. This table is the result of accumulated experience.

The seismic acceleration in the vertical direction is less important compared to the acceleration in the horizontal direction. Structures can normally withstand these vertical forces. However, in order to allow for the increase in gravity forces due to the vertical acceleration, the permissible soil pressure is limited

The most destructive force is caused by horizontal earth motion. When the ground underneath a structure is moved suddenly to one side, the building will tend to remain in its original position because of its inertia. The acceleration of the horizontal movement varies and its maximum value is the yardstick commonly adopted for measuring the equivalent static earthquake force. If the acceleration of the horizontal earth movement is one-tenth the acceleration due to gravity (Rossi-Forel Scale 9), it is assumed that the stresses in the structure caused by the earthquake are the same as those produced by horizontal static forces equal to one-tenth of the gravity forces acting on the building. While other methods of evaluating earthquake forces have been proposed and used, the one described is specified in many building codes as a practical means of safe design. It is recognised that it is not an exact measure of the effect of the chaotic ground movements that actually occur. Buildings designed on this basis have however satisfactorily resisted earthquakes.

An earthquake may occur in any direction; so buildings should be able to resist lateral forces in any direction. The earth movement, however, can be replaced by two components acting parallel to the axis of the building; hence, it is sufficient to investigate its strength in two perpendicular directions.

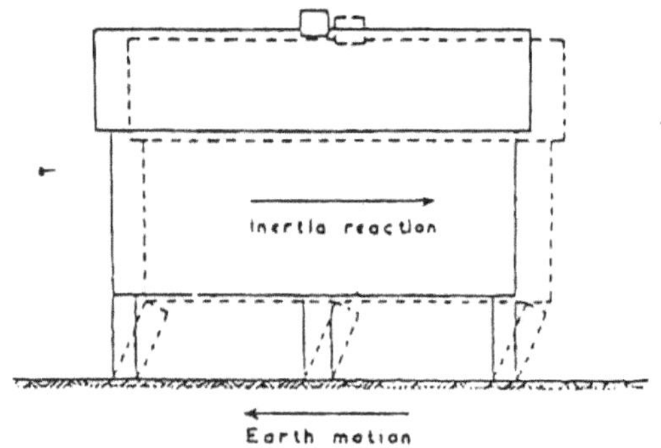

Fig. I Effect of earthquake on a building

To illustrate the effect of earthquakes on buildings, *Fig. I* shows a house set above the ground on piers. The bottoms of these piers must move with the ground and the tendency is to transmit the motion to the building above. The sudden change from rest to motion is the cause of forces (inertia reactions) acting on the superstructure in a direction opposite to that of the earth movement. To make provision for the stresses caused by these forces is the purpose of the earthquake design.

A building must be designed to withstand the horizontal seismic acceleration at any plane. In order to obtain the horizontal acceleration, the mass or dead load above the plane under consideration is taken into account and also the effective live load above the plane. When the horizontal forces have been decided upon, the stresses can be computed in a manner similar to the determination of wind stresses. In multi-storeyed buildings a certain percentage of the mass at each floor is assumed to act at that level in a horizontal direction. Contrary to wind stress analysis, all parts of a building, exterior as well as interior, are subjected to the lateral earthquake forces.

As earthquake produces movements in the ground in the form of waves it would be reasonable to take into account the "ridge" and "valley" effects in buildings similar to the "hogging" and "sagging" effects in ships but fortunately the wave amplitude (namely, the maximum distance of travel of a particle from the centre of its path) of the main wave is so large as to ensure against this possibility.

Every building has a natural period of vibration. When vibrations are introduced in a building by an earthquake having the same period of vibration as the natural period of vibration of the building, resonance would occur, namely, the vibrations would be intensely magnified and the building would be completely destroyed. The vibration periods of earthquakes have a rate of I to 2·5 seconds. The self

vibration periods of reinforced concrete buildings vary between 0·3 and 0·5 second for buildings upto 27·50m in height; for greater heights the vibration periods may be similar to those of earthquakes. Possibilities of resonance therefore exist with reinforced concrete buildings higher than 27·50m and greater heights should therefore be avoided.

All parts of a building should be firmly tied together and so stiffly braced that the building will tend to move as a unit. Floors and cross walls should be continuous throughout the building and openings should, as far as possible, be placed away from outside corners. Symmetry in the arrangement of cross walls or approximate coincidence of the centre of mass and the centre of rigidity (the point where a lateral force must be placed to produce equal deflections of the cross walls) is desirable: otherwise forces due to rotation must be taken into consideration.

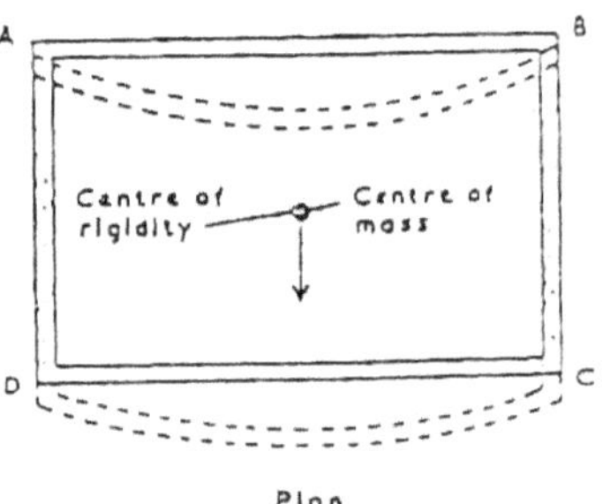

(a) Equal deflections of walls

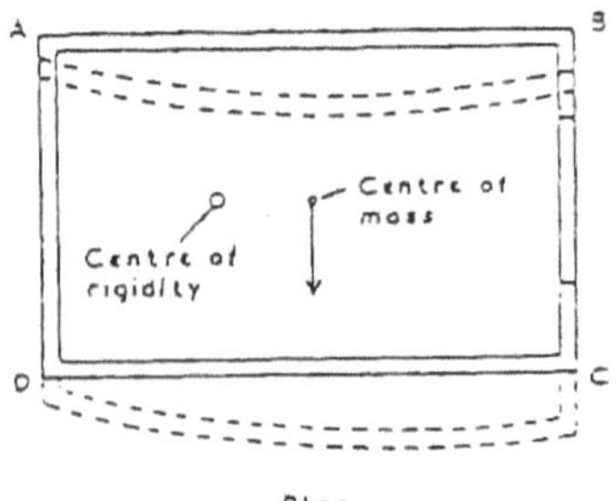

(b) Unequal deflections of walls

Rossi-Forel scale	Horizontal acceleration	REMARKS
1	0·0005g	Almost imperceptible, being felt only by an experienced observer.
2	0·001g	Feeble, being felt by a small number of persons at rest.
3	0·0025g	Very slight, being felt by several persons at rest.
4	0·006g	Slight, being felt by several persons in motion and resulting in disturbance of movable objects, doors, windows, etc.
5	0·01g	Weak, being felt generally by every one and resulting in disturbance of furniture, etc.
6	0·0125g	Moderate, resulting in general awakening of those asleep and disturbance of trees and shrubs
7	0·025g	Strong, resulting in fall of plaster; no serious damage to buildings but general panic.
8	0·05g	Very strong, resulting in cracks in walls of buildings, fall of chimneys, etc.
9	0·1g	Severe, resulting in some buildings being partially or wholly destroyed.
10	0·3g	Violent, destructive, resulting in disasters, landslides, cracks in ground, etc.

RESPONSE OF STRUCTURE TO SEISMIC FORCES.

Earthquakes cause random motion of ground which can be resolved in any three mutually perpendicular directions. This motion causes structure to vibrate and the predominant vibration is HORIZONTAL.

The response of the structure to the ground vibration is found to be a function of

(i) Nature of foundation soil

(ii) Materials, form, size and mode or method or type of construction of the structure.

(iii) Duration and intensity of ground motion.

(a) In the case of structures designed for HORIZONTAL SEISMIC FORCE only, it shall be considered to act in any one direction at a time so as to give the worst effect. The building as a rule should be checked for seismic forces in both the directions separately

(b) When both horizontal and vertical seismic forces are considered in the design HORIZONTAL FORCE in any one direction at a time may be considered simultaneously with the vertical force which may be taken as HALF of the HORIZONTAL seismic force.

However the VERTICAL seismic force shall be considered in the case of structures in which STABILITY is a criterion of design or for overall STABILITY is a criterion of design OR for overall stability of analysis of structure.

ASSUMPTIONS

(i) Earthquakes cause impulsive ground motion which is complex and irregular in character, changing in period and amplitude and

Lasting for a small duration and thus SINUSOIDAL excitations will not occur as it would need time to build up such amplitudes, under resonance of this type. as Visualised under steady state.

(ii) An earthquake is not likely to occur simultaneously with wind or floods.

(iii) The structure is assumed to stand on a soil which will not settle or slide due to Vibration Lasting for a few seconds appreciably.

(v) DL + parts of LL as would be imposed on the structure during the period of erection + Wind or Saismic loads + erection Loads.

However, For design purposes Wind load and Seismic loads shall be assumed not to act Simultaneously. Both forces shall be investigated separately and adequately provided in the design for the effects Caused by these forces taking into account the worst possible Conditions.

LOAD COMBINATIONS

To ensure the required safety and economy in the design of structures, a judicious combination of the working loads is necessary

The specified loads should be combined in accordance with relevant Codes. However in the absence of such recommendations, the following Load Combinations may be adopted.

(i) Dead Load alone (DL)

(ii) DL + Partial or Full LL which Causes the most Critical Condition in the structure

(iii) DL + Wind or Seismic Load

(iv) DL + part of or whole of the specified LL whichever is most likely to occur in Combination with the specified Wind or Seismic Loads + Wind or Seismic Loads.

EARTHQUAKE
LOADS

EARTHQUAKE LOADS
BASE - SHEAR - CALCULATIONS.

Applicable for structures
$H/B < 3$ where

H = Total height of structure
B = Total width of structure
Lateral force due to Earthquake loads
$= F_z = \alpha \beta \gamma_z \delta \theta \eta \, w_z$
at any level z

V_B = Base shear = Total Lateral Force at the base of the structure

$$\sum_{z=1}^{z=n} F_z = \alpha \beta \delta \theta \eta \sum_{z=1}^{z=n} \gamma_z w_z$$

α = Intensity Factor (Table T-1)
β = Dynamic Factor (T-3)
γ_z = Height Factor (T-4)
δ = soil factor (T-5)
θ = Behaviour Factor (T-6)
η = Importance Factor (T-7)

w_z = DL + LL (appropriate live load) of that floor under consideration (Appropriate Live Load Table - T-8)

INTENSITY FACTOR - α

TABLE - T-1

Zone	A	B	C	D
α	0.75	0.50	0.30	0.10

FUNDAMENTAL TIME PERIOD - T

TABLE - T-2

Type or Description of structure	T
Multi-storey buildings. (Brick stone masonry or Concrete walls)	$T = \dfrac{0.06 H}{\sqrt{B}} \sqrt{\dfrac{H}{2B+H}}$
Multi-storey buildings with reinforced Concrete shear walls	$T = \dfrac{0.08 H}{\sqrt{B}} \sqrt{\dfrac{H}{B+H}}$
Multi-storey reinforced Concrete framed buildings	$T = \dfrac{0.09 H}{\sqrt{B}}$
Multi-storey steel framed buildings	$T = 0.1 \dfrac{H}{\sqrt{B}}$
other special structures	To be assesed from tests & model analysis

SEISMIC ZONE MAP OF JORDAN

DYNAMIC FACTOR - β

TABLE - T.3
Applicable for $H \leq 50m$

Type or description of structure	β
Multi-storey residential buildings	$\dfrac{0.05}{\sqrt[3]{T}}\ \begin{array}{l}\geq .04\\ \leq 0.10\end{array}$
Multi-storey buildings warehouses	$\dfrac{0.06}{\sqrt[3]{T}}\ \begin{array}{l}\geq .05\\ \leq 0.12\end{array}$
Tall structures Chimneys - Towers	$\dfrac{0.01}{\sqrt[3]{T}}\ \begin{array}{l}\geq 0.06\\ \leq 0.20\end{array}$
structures of two floors or less and other type of structures	0.10

HEIGHT FACTOR - γ_z

TABLE - T-4

Type or description of structure	γ_z
structures of two floors or less, Bridges etc.	$\gamma_z = 1.0$

Type or description of structure	γ_z
Multi-storey building with nearly equal in floor to floor heights and loading	$\gamma_z = \dfrac{3z}{2n+1}$
Multi-storey structures $\leq 50m$	$\gamma_z = \dfrac{h_z \sum\limits_{z=1}^{n} w_z h_z}{\sum\limits_{z=1}^{n} w_z h_z^2}$
structures of Considerable relative amplitude	$\gamma_z = A\,\dfrac{\sum\limits_{z=1}^{n} w_z A_z}{\sum\limits_{z=1}^{n} w_z A_z^2}$

A_z = relative amplitude at level z

A = relative amplitude

h_z = Height measured from the base of the building to the level of floor (z) under Consideration.

SOIL FACTOR - δ

TABLE - T-5

$$\delta = \dfrac{0.7}{\sqrt[3]{T-T_s}}\ \begin{array}{l}\geq 0.8 \text{ and}\\ \leq 1.3\end{array}$$

Type of Soil	T_s
Igneous, Sedimentary, Metamorphic rocks	0.2
Gravel - rock layers Firm soils - upto 15 m thick Loose sandy upto 5m .	0.4
Consolidate layers of sandy soil (15 to 80m) thick	0.4 - 0.8
Filled soils 2 - 30m thick Firm Compacted (5 - 140m)	0.4 - 1.4
Filled soils > 30m Loose sandy soils > 140m	1.4

BEHAVIOUR FACTOR - θ

TABLE - T-6

Type of structure	θ
Reinforced concrete buildings. Retaining walls Chimneys	1.0
Ductile framed (steel) structures designed to resist Lateral forces	0.67
Framed structures with shear walls and Cores of box type to resist Lateral loads	1.33
Frames & shear walls designed to resist Lateral Loads together	0.80
Water Towers	2.5

IMPORTANCE FACTOR - η

TABLE - T-7

Type or description of structure	η
Important service & Community structures Hospitals. Power houses etc.	1.3
Mosques, churches, cinema halls. - shoping centres	1.2
Residential & other types	1.0

REDUCTION FACTOR ON LIVE LOADS - K

$$W_z = G_z + K\, Q_z$$
$$G = DL$$
$$Q = LL$$

TABLE - T-8

Type or description of structures	K
Residential buildings	0
Important service and Community structures Hospitals, Fire brigade emergency etc.	0.25
Ware houses - Water Towers.	1.0

when $w_1 = w_2 = w_2 = \cdots = w_n$

$$\frac{F_1}{h_1} = \frac{F_2}{h_2} = \cdots = \frac{F_z}{h_z} = \cdots = \frac{F_n}{h_n}$$

$$V = F_1 + F_2 + F_3 + F_4 + \cdots + F_z + \cdots F_n$$

$$V = \frac{F_z}{h_z}\left(h_1 + h_2 + \cdots + h_z + \cdots h_n\right)$$

$$F_z = \frac{V\, h_z}{\sum_{z=1}^{n} h_z} \;,\quad \text{if } w_1 \neq w_2 \text{ etc.}$$

$$F_z = V\, \frac{h_z w_z}{\sum_{z=1}^{n} h_z w_z}$$

Design Session III

Flexural Analysis of Beams

Introduction :

* Assume that a small transverse load is placed on a concrete beam with tensile reinforcement and that the load is gradually increased in magnitude until the beam fails.

* The beam will go through three distinct stages before collapse occurs:
  (1) the uncracked concrete stage.
  (2) the concrete cracked-elastic stresses range.
  (3) the ultimate strength stage.

* A relatively long beam is considered so that shear will not be a factor.

---

(1) Uncracked Concrete Stage:

* modulus of rupture: bending tensile stress at which concrete begins to crack.

* At small loads when the tensile stresses are less than the modulus of rupture, the entire cross-section of the beam resists bending, with compression on one side and tension on the other.

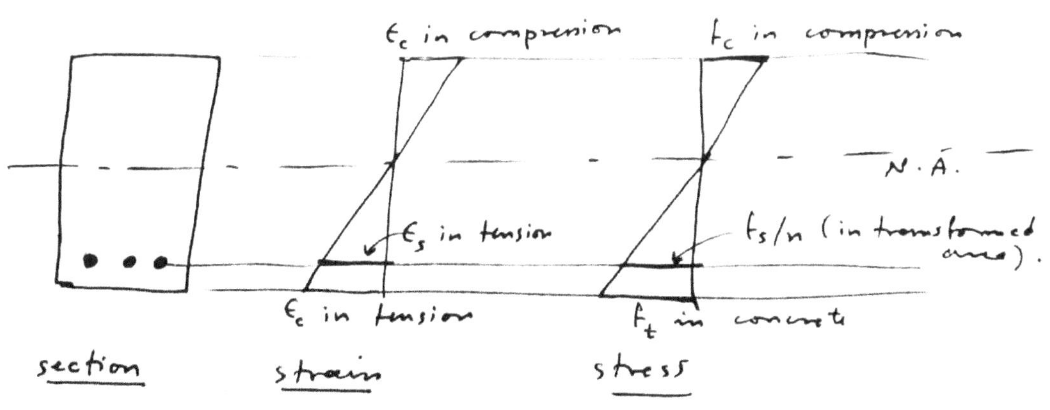

---

(2) Concrete Cracked-Elastic Stresses Stage.

* As the load is increased after the modulus of rupture of the beam is exceeded, cracks begin to develop in the bottom of the beam.

* The <u>cracking moment</u>, M_{cr}, is the moment at which these cracks begin to form, i.e., when the tensile stress in the bottom of the beam equals the modulus of rupture.

* As the load is further increased, these cracks quickly spread up to the vicinity of the neutral axis, and then the neutral axis begins to move upward.

* The <u>cracks</u> occur at those places along the beam where the <u>actual</u> moment is greater than the cracking moment.

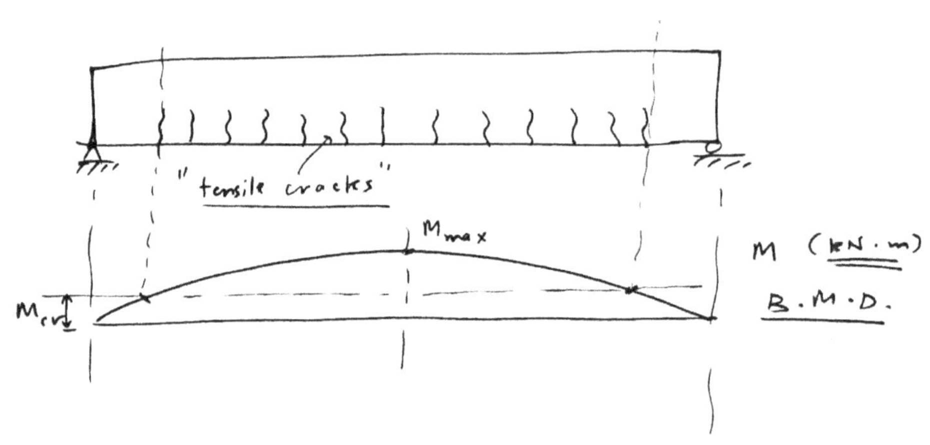

* When the bottom has cracked, another stage is present because the concrete in the cracked zone cannot resist tensile stresses — the steel must do it.

* This stage will <u>continue</u> as long as:
  1. the compression stress in the top fibers is less than $\approx 0.5 f_c'$

and, 2. the steel stress is less than its yield point, f_y.

* In this stage, the compressive stresses vary linearly with the distance from the neutral axis.

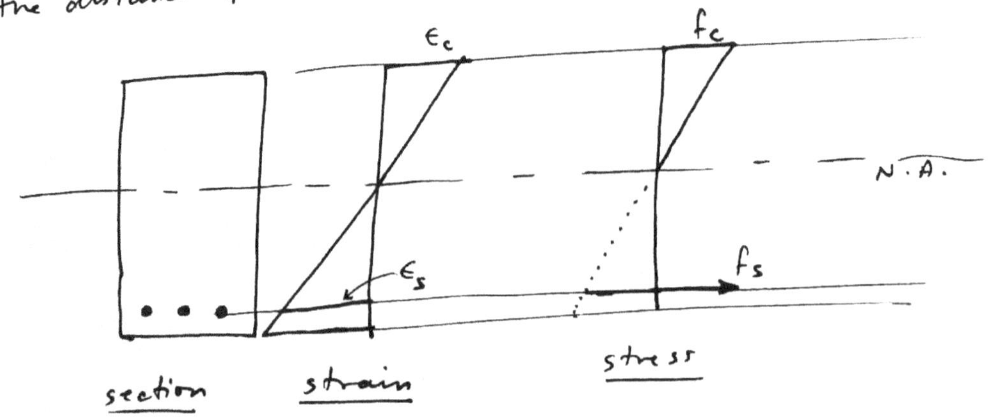

* The <u>straight line</u> stress-strain variation normally occurs in reinforced concrete beams under normal <u>service-loads</u>.

* This is because at service loads, the stresses are generally less than $0.5 f_c'$.

* To compute the concrete and steel stresses in this stage, the "<u>transformed-area method</u>" is used.

* The <u>service</u> or <u>working loads</u> are the loads which are assumed to actually occur when a structure is in use or service.

* Under these loads, moments develop which are considerably larger than the cracking moment.

* In this stage, the tensile side of the beam will be cracked. Later, we will learn to estimate crack widths and methods of limiting their widths.

---

③ <u>Beam Failure — Ultimate Strength Stage</u>:

* As the load is increased further so that the compressive stresses are greater than $0.5 f_c'$, the tensile cracks move further upward, as does the neutral axis, and the concrete stresses begin to change appreciably from a straight line.

* For this initial discussion, it is assumed that the reinforcing bars have yielded.

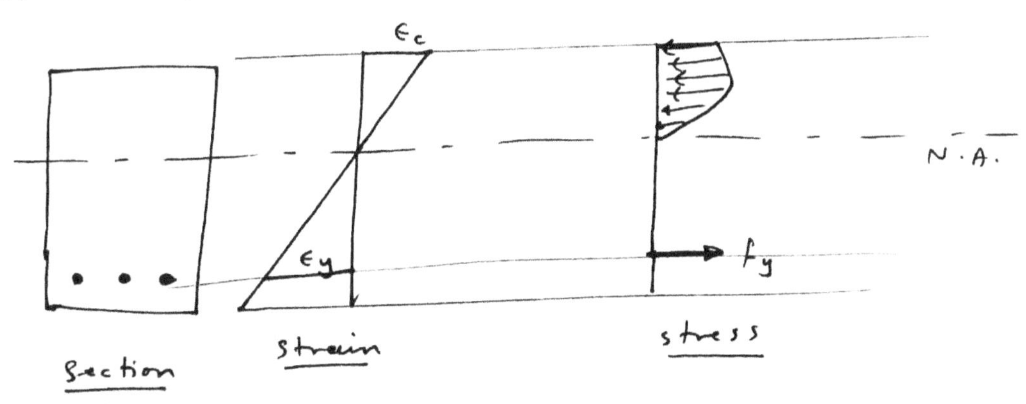

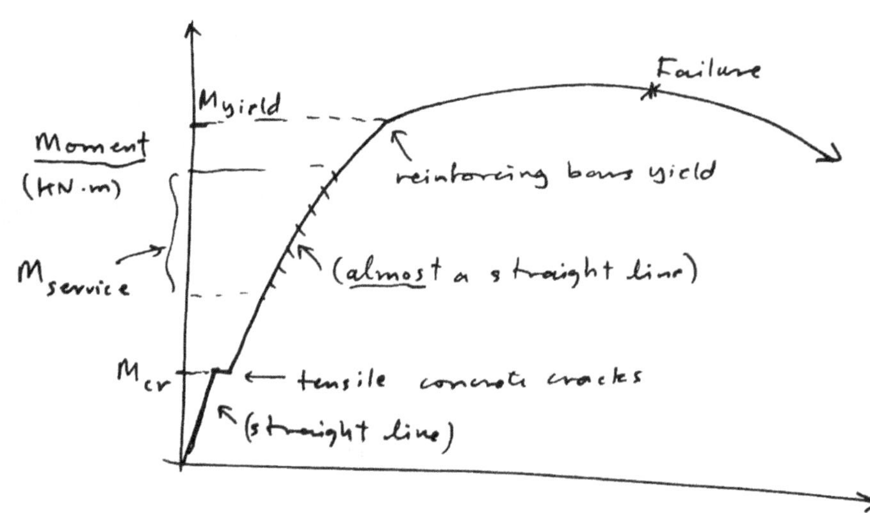

$$\text{Curvature } \theta = \frac{\epsilon}{y}.$$

---

1 Cracking Moment, M_{cr} :

* The area of reinforcement as a percentage of the total cross-sectional area of a beam is quite small (usually 2% or less), and its effect on the beam properties is almost negligible as long as the beam is <u>uncracked</u>.

* In this stage (uncracked concrete), the calculation of bending stresses in the beam can be based on the <u>gross</u> properties of the beam's cross-section.

$$f = \frac{My}{I_g} \quad , \quad I_g : \text{gross moment of inertia of the section.}$$

* The steel area A_s is neglected compared with the concrete area.
* The cracking moment, M_{cr}, is calculated from the above formula when the tensile stress f equals the modulus of rupture, f_r.

$$f_r = \frac{M_{cr} \cdot c}{I_g}$$

$$\Rightarrow \quad M_{cr} = \frac{f_r \, I_g}{c} \, .$$

Example 1 :

(a) Assuming the concrete is uncracked, compute the bending stresses in the extreme fibers of the beam shown in the figure for a bending moment of 32 kN·m. The concrete has an $f_c' = 30$ MPa.

(b) Determine the cracking moment of the section.

Solution :

(a) $I_g = \frac{1}{12} b h^3$

$\quad = \frac{1}{12} (300)(450)^3$

$\quad = 2278 \times 10^6$ mm^4.

Bending stress $f = \frac{My}{I_g}$

$f_{max.} = \frac{Mc}{I_g} = \frac{(32)(225 \times 10^{-3})}{2278 \times 10^6 \times 10^{-12}}$

$\quad = 3160$ kN/m^2

$\quad = 3.16$ MPa.

$c = \frac{450}{2} = 225$ mm

[Note : 1 MPa = 1000 kN/m^2].

tensile strength of concrete (in bending) $\equiv$ modulus of rupture

$\equiv f_r = 0.62 \sqrt{f_c'}$, f_c' in MPa.

$\quad = 0.62 \sqrt{30}$

$\quad = 3.396$ MPa.

Since $f = 3.16$ MPa $< f_r = 3.396$ MPa, the section is assumed uncracked, correctly.

(b) Compute the cracking moment, M_{cr}, based on $f = f_r$.

$f = f_r = \frac{M_{cr} \cdot c}{I_g}$

$3.396 \times 1000 = \frac{M_{cr} \cdot (225 \times 10^{-3})}{2278 \times 10^6 \times 10^{-12}}$

$\Rightarrow M_{cr} = 34.38$ kN·m

(max. moment allowed before tensile cracks occur).

② Elastic Stresses — Concrete Cracked:

* When the bending moment is sufficiently large to cause the tensile stress in the extreme fibers to be greater than the modulus of rupture, it is assumed that __all__ of the concrete on the tensile side of the beam is __cracked__, and must be neglected in the flexure calculations.

* The cracking moment of a beam is normally quite small compared to the service load moment. Thus, when the service loads are applied, the bottom of the beam cracks.

* The cracking of the beam does __not__ necessarily mean that the beam is going to fail. The reinforcing bars on the tensile side begin to pick up the tension caused by the applied moment.

* On the tension side of the beam an assumption of __perfect bond__ is made between the reinforcing bars and the concrete.

* Define the __modular ratio__ $n = \dfrac{E_s}{E_c}$.

* Strain in Steel = Strain in Concrete (__Compatibility__).

$$\varepsilon_s = \varepsilon_c \quad , \quad but \quad f = E\varepsilon$$

$$\frac{f_s}{E_s} = \frac{f_t}{E_c} \implies f_s = \left(\frac{E_s}{E_c}\right) f_t = n f_t \, .$$

$$\implies f_t = \frac{f_s}{n} \, .$$

* Replace the steel bars with an equivalent area of fictitious concrete $(\underline{nA_s})$, which supposedly can resist tension.

* This area is referred to as the __transformed area__, A_t.

$$A_t = n A_s \, .$$

* The __transformed section__ can be analyzed by the usual methods of elastic homogeneous beams.

* The concrete is assumed to be cracked and unable to resist tension.

* $f_t = f_s/n$ is the fictitious stress in the concrete if it could carry tension.

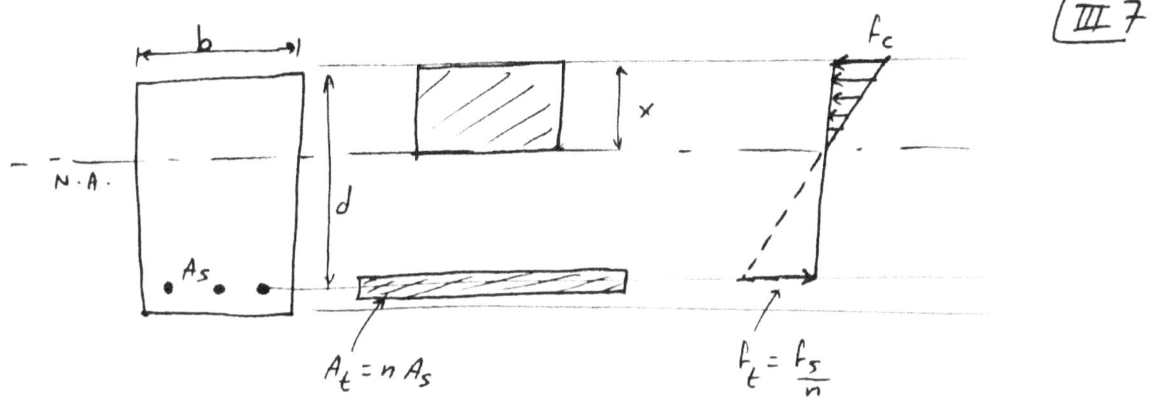

* <u>First Step</u> in the <u>transformed-area</u> method :

  Locate the neutral axis (i.e find x ?).

* Moment of Compression Area about N.A. = Moment of Tensile Area about N.A.

* You will get a quadratic equation in x.

* <u>Second Step</u> is to calculate I of the transformed section.

* <u>Third Step</u> is to calculate the stresses using the flexure

    formula $f = My / I$.

<u>Example 2</u> :

  Calculate the bending stresses in the beam shown in the figure by using the transformed area method. Take $n = 9$ and

  $M = 93 \ kN \cdot m$.

<u>Solution</u> :

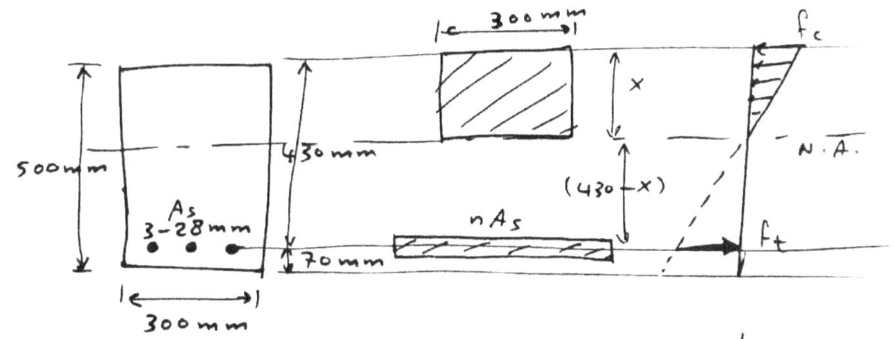

$$A_t = n A_s = 9(1846) = 16614 \ mm^2.$$

<u>Step 1</u> : Locate the neutral axis (Find x ?) :

$$300 x \left(\frac{x}{2}\right) = n A_s (430 - x) \ , \quad \text{where} \quad A_s = 3 \left[\frac{\pi (28)^2}{4} \right] = 1846 \ mm^2$$

$$150 x^2 = 9(1846)(430 - x)$$

$$x^2 = 110.76 (430 - x)$$

$$x^2 = 47626.8 - 110.76 x$$

$$x^2 + 110.76x - 47626.8 = 0$$

$$x = \frac{-110.76 \pm \sqrt{(110.76)^2 - 4(1)(-47626.8)}}{2(1)} = 170 \ mm.$$

Step 2 :
$$I = \frac{1}{3}(300)(170)^3 + (16614)(430-170)^2$$
$$= 1614.41 \times 10^6 \ mm^4.$$

Step 3 :
$$f_c = \frac{My}{I} = \frac{(93)(170 \times 10^{-3})}{1614.41 \times 10^6 \times 10^{-12}} = 9790 \ kN/m^2$$
$$= 9.79 \ MPa.$$

$$f_t = \frac{My}{I} = \frac{(93)(430-170) \times 10^{-3}}{1614.41 \times 10^6 \times 10^{-12}} = 14980 \ kN/m^2$$
$$= 14.98 \ MPa.$$

$$\text{but} \ \ f_t = \frac{f_s}{n} \implies f_s = n f_t = 9(14.98) = 135 \ MPa.$$

Example 3 :

Determine the <u>allowable resisting moment</u> of the beam of Example 2 if the allowable stresses are $f_c = 9.3$ MPa and $f_s = 140$ MPa.

Solution :

$$f_c = \frac{My}{I} \implies 9.3 \times 1000 = \frac{M(170 \times 10^{-3})}{1614.41 \times 10^{-6}}$$

$$\implies M = \underline{\underline{88.3}} \ kN \cdot m.$$

$$f_s = n f_t = n\frac{My}{I} \implies 140 \times 1000 = \frac{9M(430-170) \times 10^{-3}}{(1614.41 \times 10^{-6})}$$

$$\implies M = \underline{\underline{96.6}} \ kN \cdot m.$$

<u>Note</u>: For a given beam, the concrete and steel will not usually reach their maximum allowable stresses at exactly the same bending moments.

The resisting moment of this section $M = 88.3$ kN·m. (when concrete becomes overstressed).

Example 4 :

Compute the bending stresses in the beam shown in the figure by using the transformed area method. Take $n = 8$ and $M = 145$ kN·m.

Solution :

$$A_S = 4\left[\frac{\pi (32)^2}{4}\right]$$

$$= 3215 \text{ mm}^2 .$$

$$A_t = n A_s = 8(3215)$$

$$= 25{,}720 \text{ mm}^2 .$$

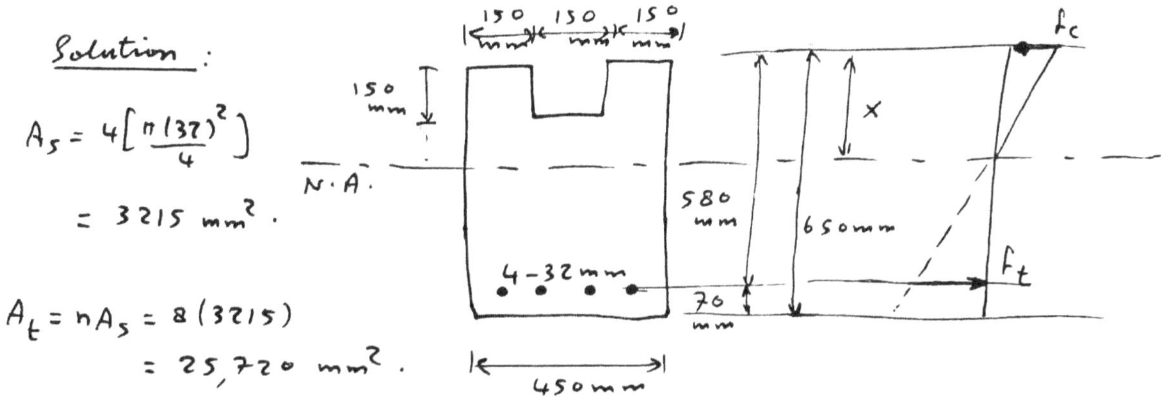

Step 1 : Locate the neutral axis (find x ?) :

[Assume N.A. is below the hole].

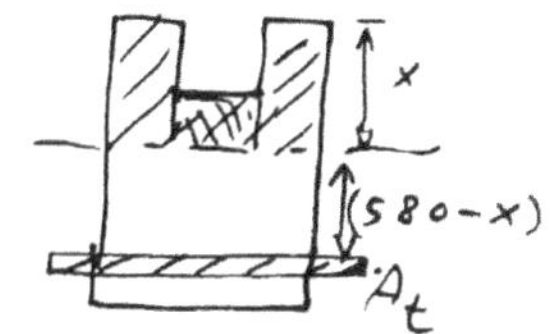

$$450 \times \left(\frac{x}{2}\right) - (150)(150)\left(x - \frac{150}{2}\right) = 25{,}720\,(580 - x)$$

$$225 x^2 - 22{,}500 x + 1{,}687{,}500 = 14{,}917{,}600 - 25{,}720 x$$

$$225 x^2 + 3220 x - 13{,}230{,}100 = 0$$

$$x = \frac{-3220 \pm \sqrt{(3220)^2 - 4(225)(-13{,}230{,}100)}}{2(225)} = 235 \text{ mm}.$$

$$\therefore \quad x = 235 \text{ mm} > 150 \text{ mm} .$$

$\therefore$ Our assumption is correct (N.A. is below the hole).

Step 2 : Calculate I :

$$I = 2\left[\tfrac{1}{3}(150)(235)^3\right] + \tfrac{1}{3}(150)(235 - 150)^3 + 25{,}720\,(580 - 235)^2$$

$$= \underset{\sim}{\;} 4389 \times 10^6 \text{ mm}^4 .$$

Step 3 : Compute stresses :

$$f_c = \frac{My}{I} = \frac{(145)(235 \times 10^{-3})}{4360.339 \times 10^6 \times 10^{-12}} = 7762 \text{ kN/m}^2$$

$$= 7.81 \text{ MPa} .$$

$$f_s = n f_t = n \frac{My}{I} = (8) \frac{(145)(580 - 235) \times 10^{-3}}{4360.339 \times 10^6 \times 10^{-12}}$$

$$= 91182 \quad kN/m^2$$

$$= 91.182 \quad MPa.$$

---

* A __doubly-reinforced__ concrete beam has __both__ compression steel and tensile steel.

* __Advantages__ of compression steel :

    1. permits the use of smaller beams.

    2. reduced sizes can be very important when space or architectural requirements limit the sizes of beams.

    3. it is quite helpful in reducing long-term deflections

    4. it is useful for positioning stirrups or shear reinforcement

* __Disadvantages__ of compression steel :

    1. it is uneconomical.

---

* A_s' : area of compression reinforcement.

    transformed area $= 2n A_s'$.

    Actually, we should subtract the steel area, A_s'.

    $\therefore$ Actual transformed area $= 2n A_s' - A_s' = (2n-1) A_s'$.
      ($+$ concrete compression area).

__Example 5__ :

    Compute the bending stresses in the beam shown in the figure. Take $n = 10$ and $M = 157$ kN·m.

__Solution__ :

$$A_s = 4\left[\frac{\pi(28)^2}{4}\right] = 2462 \ mm^2.$$

$$A_s' = 2\left[\frac{\pi(28)^2}{4}\right] = 1231 \ mm^2.$$

$$A_t = n A_s = 10\,(2462)$$
$$= 24{,}620 \ mm^2.$$

$$A_t' = (2n-1)A_s' = 19\,(1231) = 23{,}387 \ mm^2.$$

Step 1 : Locate the N.A. (Find x ?)

$$350 x \left(\frac{x}{2}\right) + 23,387 (x-60) = 24,620 (380+60-x)$$

$$175 x^2 + 23,387 x - 1,403,220 = 10,832,800 - 24,620 x$$

$$175 x^2 + 48,007 x - 12,236,020 = 0$$

$$x = \frac{-48,007 \pm \sqrt{(48,007)^2 - 4(175)(-12,236,020)}}{2(175)} = 161 \text{ mm}.$$

Step 2 : Calculate I :

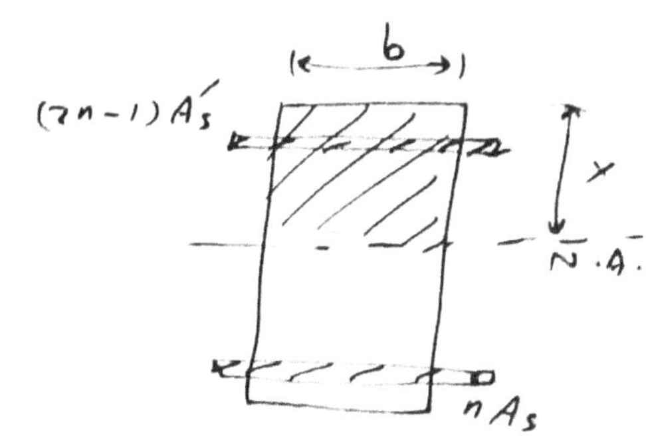

$$I = \frac{1}{3} (350)(161)^3 + 23,387 (161-60)^2$$
$$+ 24,620 (380+60-161)^2$$

$$= 2641.899 \times 10^6 \text{ mm}^4.$$

Step 3 : Calculate the bending stresses:

$$f_c = \frac{My}{I} = \frac{(157)(161 \times 10^{-3})}{2641.899 \times 10^6 \times 10^{-12}} = 9570 \text{ kN/m}^2$$

$$= 9.57 \text{ MPa}.$$

$$f_s = n f_t = n \frac{My}{I} = \frac{(10)(157)(380+60-161)\times 10^{-3}}{2641.899 \times 10^6 \times 10^{-12}} = 165,800 \text{ kN/m}^2$$

$$= 165.8 \text{ MPa}.$$
$$\approx 166 \text{ MPa}.$$

$$f_s' = 2n f_t' = 2n \frac{My}{I} = 2(10) \frac{(157)(161-60)\times 10^{-3}}{2641.899 \times 10^6 \times 10^{-12}} = 120040 \text{ kN/m}^2$$

$$= 120 \text{ MPa}.$$

3 Ultimate or Nominal Flexural Moments:

* At this stage, the tensile reinforcing bars are stressed to their yield point before the concrete on the compressive side of the beam crushes.

* After the concrete compression stresses exceed about $0.5 f_c'$, they no longer vary directly as the distance from the neutral axis or as a straight line.

* The stresses vary in an approximately parabolic distribution.

* It is assumed that the curved compression diagram is replaced with a rectangular one (Whitney block) with an average stress of $0.85 f_c'$.

* Experimental tests show that with the assumptions used here, accurate flexural strengths are determined.

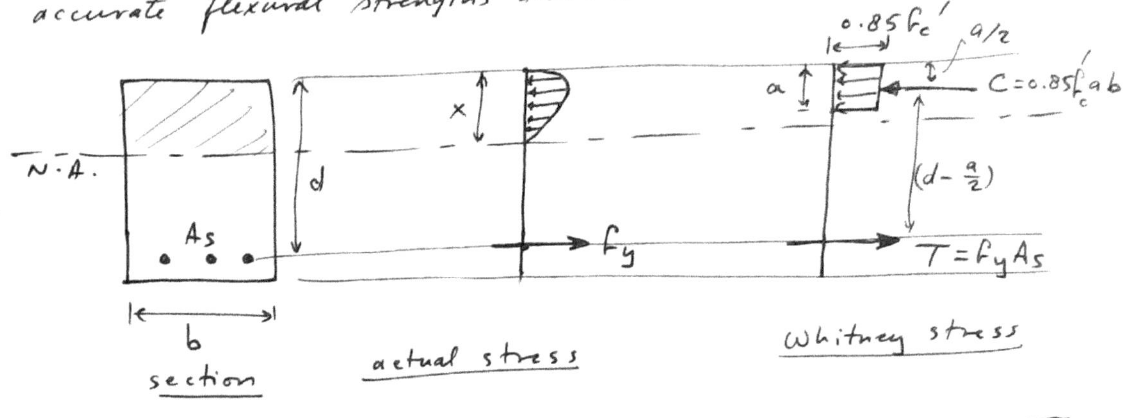

Example 6 :

Determine the nominal (theoretical ultimate) moment strength of the beam section shown in the figure if $f_y = 415$ MPa and $f_c' = 20$ MPa.

Solution:

$$A_s = 3\left[\frac{\pi (28)^2}{4}\right] = 1846 \text{ mm}^2.$$

$$C = T$$

$$0.85 f_c' a b = f_y A_s$$

$$0.85 (20) a (350) = 415 (1846)$$

$$\Rightarrow a = 129 \text{ mm}.$$

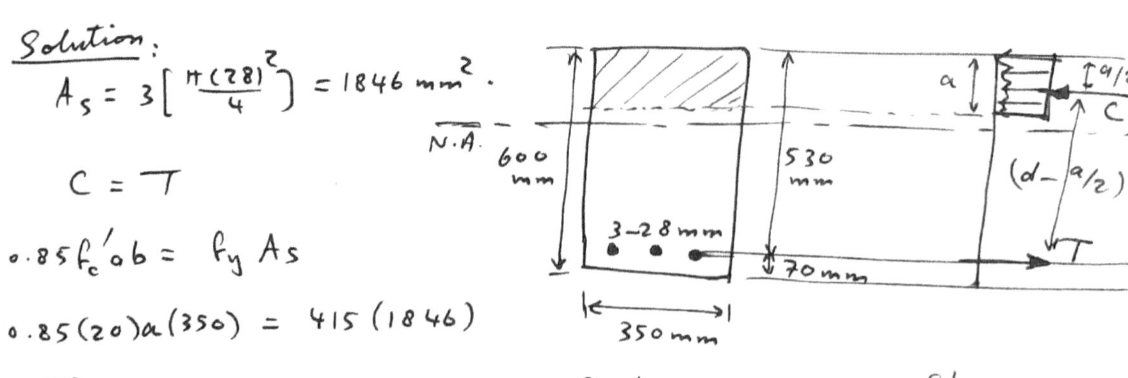

$$\therefore \quad M_n = C\left(d - \frac{a}{2}\right) \equiv T\left(d - \frac{a}{2}\right)$$

$$= (415 \times 1000)(1846 \times 10^{-6})\left(\frac{530}{1000} - \frac{1}{2} \cdot \frac{129}{1000}\right)$$

$$= 360 \ kN \cdot m.$$

Example 7 :

Calculate the nominal (theoretical ultimate) moment strength of the beam section shown in the figure if $f_y = 415$ MPa and $f_c' = 20$ MPa. The 150-mm wide ledges on top are needed for the support of precast concrete slabs.

Solution :

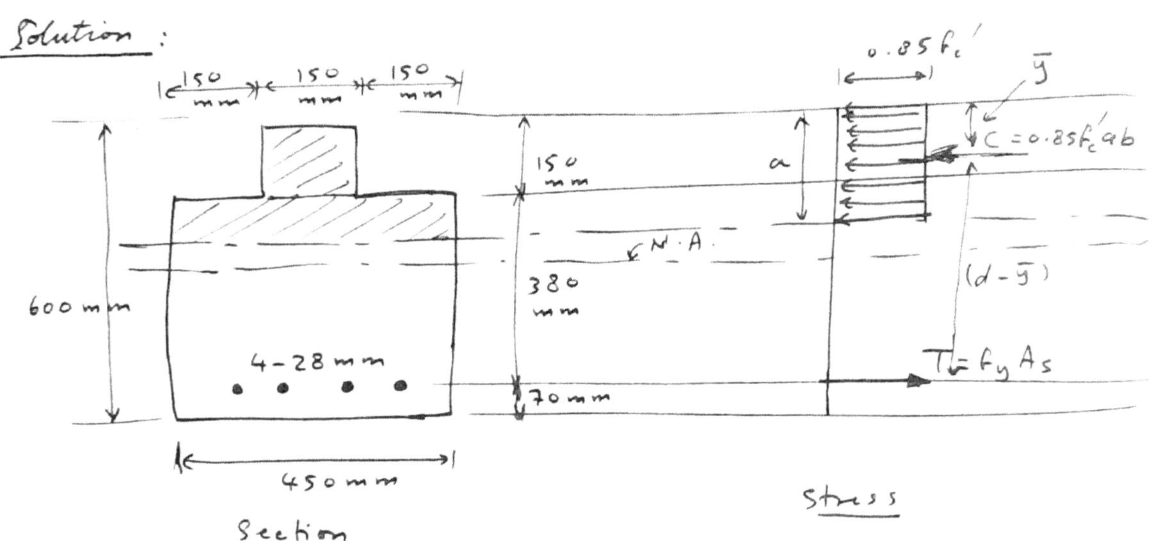

$$A_s = 4\left[\frac{\pi (28)^2}{4}\right] = 2462 \ mm^2.$$

$$T = f_y A_s \qquad and \qquad C = 0.85 f_c' A_c \ , \quad where \ A_c \equiv shaded \ area.$$

$$C = T$$

$$0.85 f_c' A_c = f_y A_s$$

$$0.85 (20) A_c = (415)(2462) \implies A_c = 60,102 \ mm^2.$$

$$But \quad A_c = (150)(150) + (450)(a - 150) = 60,102$$

$$\implies a = 234 \ mm$$

Note : C acts at the centroid $\bar{y}$ of the area A_c.

$$\bar{y} = \frac{\Sigma A\tilde{y}}{\Sigma A} = \frac{(150)(150)\left(\frac{150}{2}\right) + (450)(234 - 150)\left[150 + \left(\frac{234 - 150}{2}\right)\right]}{60,102}$$

$$\therefore \ \bar{y} = 149 \ mm.$$

$$M_n = T(d - \bar{y}) = C(d - \bar{y})$$

$$= f_y A_s (d - \bar{y})$$

$$= (415 \times 1000)(2462 \times 10^{-6})\left(\frac{380}{1000} + \frac{150}{1000} - \frac{149}{1000}\right)$$

$$= 389 \ kN \cdot m.$$

Analysis and Design of T-Beams
[Chapter 9 in Textbook]

① T Beams :

* Reinforced concrete floor systems normally consist of slabs and beams that are placed monolithically.

* Therefore, the two parts act together to resist loads.

* In effect, the beams have extra widths at their tops, called _flanges_, and the resulting T-shaped beams are called _T beams_.

* The part of a T beam below the slab is referred to as the _web_ or _stem_.

* The beams may be _L-shaped_ if the stem is at the end of a slab.

* The stirrups in the webs extend up into the slabs with the result that they further make the beams and slabs act together.

* The hatched area in the figure shows the effective size of a T beam.

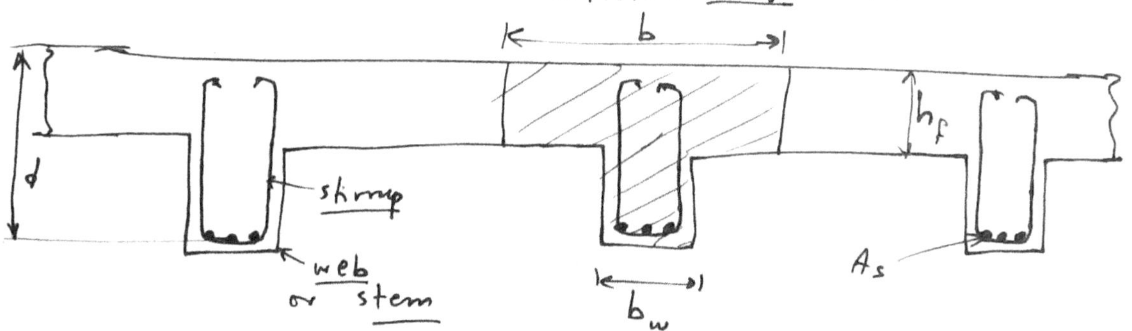

* For symmetrical T beams, the ACI Code states that the effective flange width may not exceed one-fourth of the beam span, and the overhanging width on each side may not exceed eight times the slab thickness or one-half the clear distance to the adjacent T beam.

* Isolated T-beams must have a flange thickness no less than $\frac{1}{2}$ the web width and its effective flange width may not be larger than 4 times the web width (ACI/ Section 8.10.4).

* $\rho_{max} = 0.75 \, \rho_b$.

---

* The <u>neutral axis</u> for T beams can fall either in the flange or in the web, depending on the proportion of the slabs and stems.

* If it falls in the flange, the rectangular beam formulas apply. The concrete below the neutral axis is assumed to be cracked and its shape has no effect on the flexure calculations (other than weight).

* The section above the neutral axis is rectangular.

* If the neutral axis is below the flange, the compression concrete above the neutral axis no longer consists of a single rectangle and thus the normal rectangular beam expressions do not apply.

* $a = \beta_1 x$

If the computed value of a is equal to or less than the flange thickness, the section for all practical purposes can be assumed to be rectangular even though the computed value of x is actually greater than the flange thickness.

Note : A beam does <u>not</u> have to look like a
  T beam to be one.

But the shapes, sizes and weights of the tensile concrete
do affect the deflections that occur, and their dead
weights affect the magnitudes of the moments to be
resisted.

---

(2) Analysis of T Beams :
  * Calculate the design strength ϕM_n of T beams.

Example 1 :
   Determine the permissible design strength of the T beam
shown in the figure, with $f_c' = 20$ MPa and $f_y = 350$ MPa.

Solution :

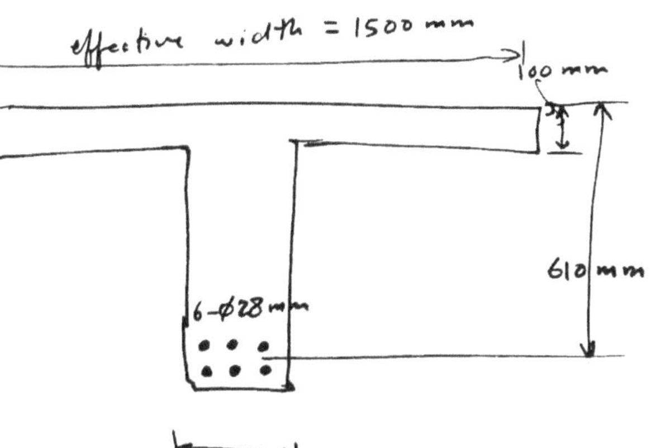

$A_s = 6\left[\dfrac{\pi(28)^2}{4}\right] = 3695$ mm^2.

$$C = T$$

$$\boxed{0.85 f_c' A_c = f_y A_s}$$

$0.85(20) A_c = (350)(3695)$

$\Rightarrow A_c = 76,074$ mm^2.

Area of Flange $= 100 \times 1500 = 150,000$ mm^2.

  Since $A_c <$ Area of Flange
  $\Rightarrow$ the stress block is entirely within the flange and
the rectangular formulas apply.

  $\Rightarrow A_c = ab \Rightarrow 76074 = a(1500)$
  $\Rightarrow a = 50.7$ mm.

Nominal Strength $M_n = T\left(d - \frac{a}{2}\right) = f_y A_s \left(d - \frac{a}{2}\right)$

$$= (350 \times 1000)(3695 \times 10^{-6})\left(\frac{610}{1000} - \frac{50.7}{2(1000)}\right)$$

$$= 756 \ kN \cdot m.$$

Design Strength $\phi M_n = (0.9)(756) = \underline{\underline{680}} \ kN \cdot m.$

* Check $A_s \leq A_{s(max)}$:

Consider the balanced-strain condition :

$$\epsilon_s = \epsilon_y = \frac{f_y}{E} = \frac{350}{200,000} = 0.00175 \ .$$

By similar triangles :

$$\frac{x_b}{0.003} = \frac{610 - x_b}{0.00175}$$

$$0.00175 \, x_b = 0.003 (610) - 0.003 \, x_b$$

$$\implies x_b = 385 \ mm > 100 \ mm .$$

$\therefore$ Neutral axis falls in the web.

$$a_b = \beta_1 x_b = 0.85 (385) = 327 \ mm > 100 \ mm .$$

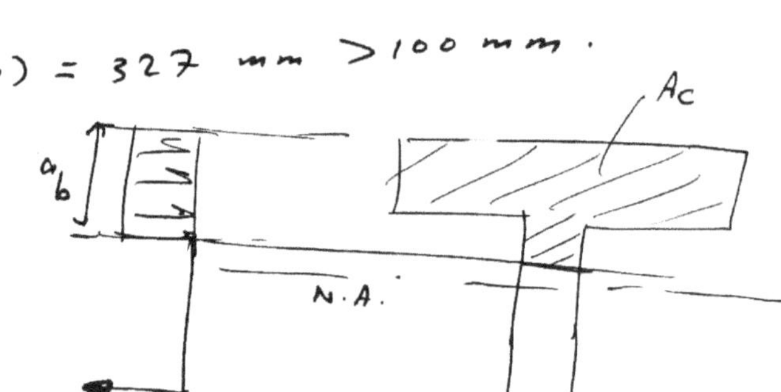

$A_c = (1500)(100)$

$\qquad + (327 - 100)(250)$

$\qquad = 206,750 \ mm^2 .$

$$C_b = T_b$$

$$0.85 f_c' A_{c_b} = f_y A_{sb}$$

$$0.85 (20)(206,750) = 350 \, A_{sb}$$

$$\implies A_{sb} = 10,042 \ mm^2 .$$

$$A_{s_{max}} = 0.75 \, A_{sb} = 0.75 (10,042) = 7532 \ mm^2 .$$

But $A_s = 3695 \ mm^2 < 7532 \ mm^2 .$

Section satisfies ACI Code Requirements.

Example 2 :

Compute the design strength for the T beam shown in the figure with $f_c' = 20$ MPa and $f_y = 350$ MPa.

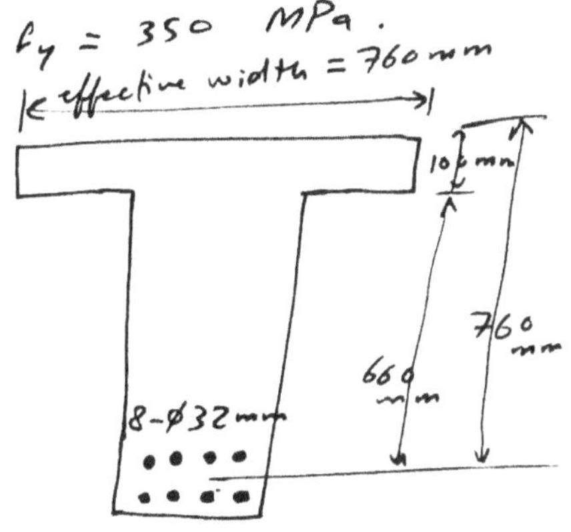

Solution :

$$A_s = 8\left[\frac{\pi(32)^2}{4}\right] = 6434 \text{ mm}^2 .$$

$$C = T$$

$$0.85 f_c' A_c = f_y A_s$$

$$0.85 (20) A_c = (350)(6434)$$

$$\implies A_c = 132,465 \text{ mm}^2 .$$

Area of Flange = $100 \times 760 = 76,000$ mm^2 .

$$\therefore \quad A_c > \text{Area of Flange}.$$

$\implies$ Stress block extends below the flange to provide the necessary compression area .

$$A_c = 76,000 + (a - 100)(350) \implies a = 261 \text{ mm} .$$

Calculate the centroid $\bar{y}$ of the compression area A_c :

$$\bar{y} = \frac{(76,000)(50) + \left(132,465 - 76,000\right)\left(\frac{261 - 100}{2} + 100\right)}{132,465}$$

(from top)

$$= 106 \text{ mm} .$$

$$\therefore \text{ Nominal Strength } M_n = T(d - \bar{y})$$

$$= f_y A_s (d - \bar{y})$$

$$= (350 \times 1000)(6434 \times 10^{-6})\left[\frac{760}{1000} - \frac{106}{1000}\right]$$

$$= 1473 \text{ kN} \cdot \text{m} .$$

Design Strength $\phi M_n = (0.9)(1473) = 1325 \text{ kN} \cdot \text{m} .$

Check ACI Code Requirements: $A_s \leq A_{s(max)}$:

For the balanced-strain condition:

$$\epsilon_s = \epsilon_y = \frac{f_y}{E_s} = \frac{350}{200,000} = 0.00175$$

By similar triangles:

$$\frac{x_b}{0.003} = \frac{760 - x_b}{0.00175}$$

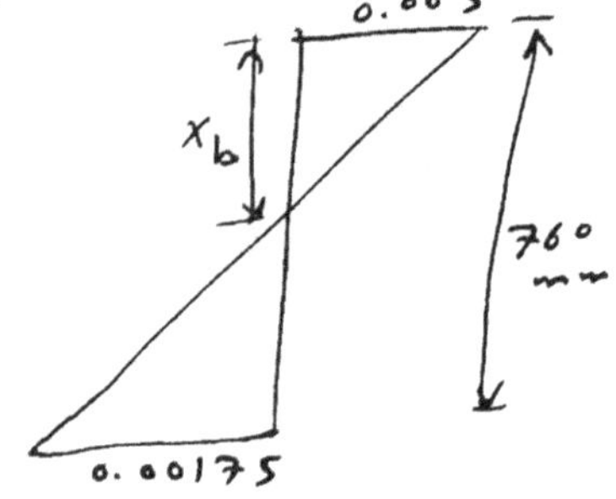

$$\Rightarrow 0.00175 x_b = 0.003(760) - 0.003 x_b$$

$$\Rightarrow x_b = 480 \text{ mm}.$$

$$x_{max} = 0.75 \, x_b = 360 \text{ mm} > 100 \text{ mm}.$$

Neutral axis falls below the flange.

$$a_{max} = 0.85 \, x_{max} = 306 \text{ mm} > 100 \text{ mm}.$$

$$A_{c(max)} = (100 \times 760) + (306 - 100)(350) = 148,100 \text{ mm}^2.$$

$$C_{max} = T_{max}$$

$$0.85 \, f_c' \, A_{c(max)} = f_y \, A_{s(max)}$$

$$0.85 (20)(148,100) = 350 \, A_{s(max)}$$

$$\Rightarrow A_{s(max)} = 7193 \text{ mm}^2.$$

But $A_s = 6434 \text{ mm}^2 < A_{s(max)} = 7193 \text{ mm}^2.$

∴ Section satisfies ACI Code requirements.

---

③ <u>Design of T Beams:</u>

* Selection of dimensions b_w, b, h_f, h:

(1) the flange dimensions are normally selected in the slab design.

(2) the web dimensions are normally selected based on shear requirements.

(3) the width of the web, b_w, is selected to satisfy beam width.

(4) Sizes are pre-selected to simplify formwork, for architectural reasons, or for deflection reasons.

---

* In the following examples, the values of d and b_w are given.

* The flanges of <u>most</u> T beams are usually so large that the neutral axis probably falls within the flange and thus the rectangular beam formulas apply.

* Should the neutral axis fall within the web, a trial-and-error process is often used for the design.

* In this process, a lever arm from the centroid of the compression block to the centroid of the steel is estimated to equal the larger of $0.9d$ or $d - \frac{1}{2}h_f$, and from this value, called z, a trial steel area is calculated.

Example 3 :

Design a T beam for the floor system shown in the figure for which b_w and d are given, where $M_D = 66$ kN·m , $M_L = 133$ kN·m , $f_c' = 28$ MPa , $f_y = 350$ mPa , and simple span $= 6$ m.

Solution :

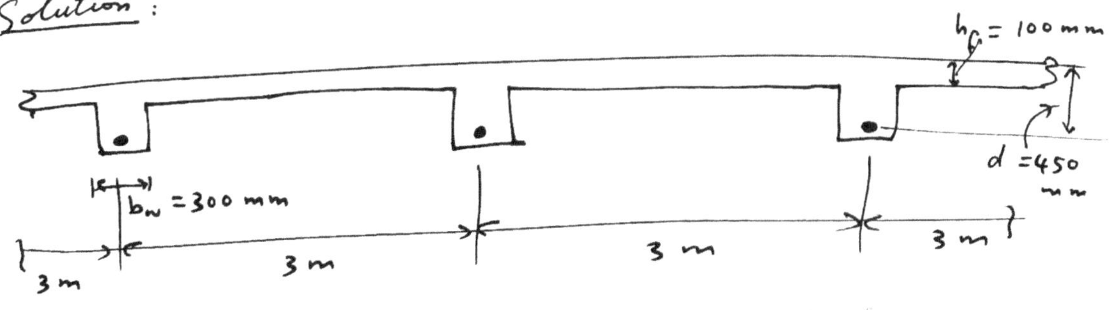

<u>Effective Flange Width</u> b :

(1) $\frac{1}{4}(6) = 1.5$ m $= 1500$ mm.

(2) $300 + (2)(8)(100) = 1900$ mm.

(3) $2\left(\frac{1}{2}\right)(3) = 3$ m $= 3000$ mm.

$\Rightarrow$ take $b = 1500$ mm.

$$M_u = 1.4\, M_D + 1.7\, M_L = 1.4(66) + 1.7(133)$$
$$= 318.5 \ \text{kN·m}$$

$$M_u \leq \phi M_n \implies M_n \geq \frac{M_u}{\phi} = \frac{318.5}{0.90} = 354 \ \text{kN·m}.$$

Assume z = larger of $\ 0.9d\ $ or $\ d - \tfrac{1}{2}h_f$:

(1) $\quad 0.9d = 0.9(450) = 405 \ \text{mm}.$

(2) $\quad d - \tfrac{1}{2}h_f = 450 - \tfrac{1}{2}(100) = 400 \ \text{mm}.$

$$\implies \text{take} \quad z = 405 \ \text{mm} \qquad (\text{lever arm}).$$

<u>Trial steel Area</u> :

$$M_n = T z = f_y A_s z$$
$$354 = (350 \times 1000)\, A_s \left(\frac{405}{1000}\right) \implies A_s = 0.002497 \ \text{m}^2 = 2497 \ \text{mm}^2.$$

$$C = T$$
$$0.85 f_c' A_c = f_y A_s$$
$$0.85(28) A_c = 350(2497) \implies A_c = 36,726 \ \text{mm}^2.$$

$$\text{Area of Flange} = b h_f = (1500)(100) = 150,000 \ \text{mm}^2.$$

$\therefore$ $A_c <$ Area of Flange.

$\therefore$ The neutral axis lies in the flange.

$$A_c = b a \implies 36,726 = 1500\, a$$
$$\implies a = 24.5 \ \text{mm}.$$

$$\therefore z = d - \frac{a}{2} = 450 - \frac{24.5}{2} = 438 \ \text{mm}.$$

Re-calculate A_s with revised z :

$$M_n = T z = f_y A_s z$$
$$354 = (350 \times 1000)\, A_s \left(\frac{438}{1000}\right)$$

$$\implies A_s = 0.002309 \ \text{m}^2 = 2309 \ \text{mm}^2.$$

$$C = T$$

$$0.85\, f_c'\, A_c = f_y\, A_s$$

$$(0.85)(28)\, A_c = (350)(2309) \implies A_c = 33{,}956 \text{ mm}^2.$$

$$A_c = ba \implies a = \frac{A_c}{b} = \frac{33{,}956}{1500} = 22.6 \text{ mm}.$$

$$\therefore \quad z = d - \frac{a}{2} = 450 - \frac{22.6}{2} = 439 \text{ mm} \approx 438 \text{ mm}.$$
$$(\text{o.k.})$$

Re-calculate A_s with revised z :

$$M_n = Tz = f_y\, A_s\, z$$

$$354 = (350 \times 1000)\, A_s \left(\frac{439}{1000}\right)$$

$$\implies A_s = 0.002304 \text{ m}^2$$
$$= 2304 \text{ mm}^2.$$

---

* Check if the section satisfies ACI code requirements,

   i.e. $\quad A_s \le A_{s(max)}$:

Consider the balanced-strain condition:

$$\mathcal{E}_s = \mathcal{E}_y = \frac{f_y}{E_s} = \frac{350}{200{,}000} = 0.00175 .$$

$$\frac{x_b}{0.003} = \frac{450 - x_b}{0.00175}$$

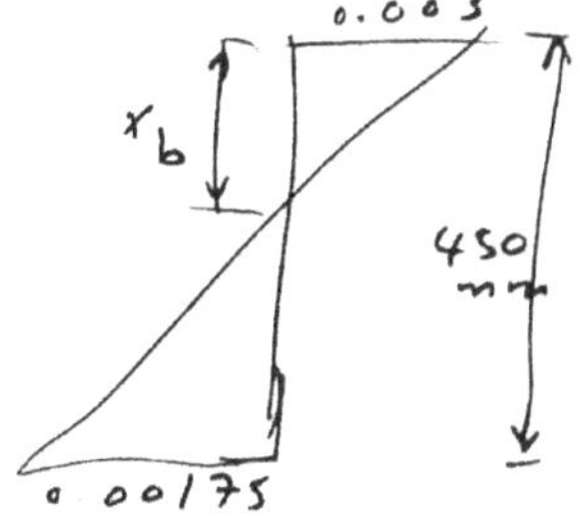

$$0.00175\, x_b = 0.003(450) - 0.003\, x_b$$

$$\implies x_b = 284 \text{ mm}.$$

$$x_{max} = 0.75\, x_b = 213 \text{ mm}.$$

$$a_{max} = \beta_1\, x_{max} = (0.85)(213) = 181 \text{ mm} > 100 \text{ mm}.$$

Neutral axis lies in the ~~flange~~ web.

$$A_{c(max)} = (100)(1500) + (181-100)(300) = 174{,}300 \text{ mm}^2.$$

$$C_{max} = T_{max}$$

$$0.85\, f_c'\, A_{c(max)} = f_y\, A_{s(max)}$$

$$0.85\,(28)\,(174{,}300) = 350\, A_{s(max)} \implies A_{s(max)} = 11{,}852 \text{ mm}^2$$

But $A_s = 2304 \ mm^2 < 11,852 \ mm^2$.

$\therefore$ Section satisfies ACI Code requirements.

* <u>check ρ_{min}</u>:

$$\rho_w = \frac{A_s}{b_w d} = \frac{2304}{300 \times 450} = 0.01707 = 1.71 \%.$$

$$\rho_{min} = \frac{1.4}{f_y} = \frac{1.4}{350} = 0.004 = 0.4\% < \rho.$$

$\therefore$ Section satisfies ACI Code requirements.

---

* Instead of calculating the maximum reinforcement area $A_{s(max)}$ for a T beam as shown above, we will derive a <u>general expression</u> for $A_{s(max)}$ in T beams:

For a balanced strain condition:

$$\epsilon_s = \epsilon_y = \frac{f_y}{E_s} = \frac{f_y}{200,000}.$$

By similar triangles:

$$\frac{x_b}{0.003} = \frac{d - x_b}{\epsilon_y}$$

$$\epsilon_y x_b = 0.003d - 0.003 x_b$$

$$(\epsilon_y + 0.003) x_b = 0.003 d$$

$$\Rightarrow \quad x_b = \frac{0.003 d}{0.003 + \epsilon_y} = \frac{0.003 d}{0.003 + \frac{f_y}{200,000}}$$

$$\Rightarrow \quad x_b = \frac{600 d}{600 + f_y} = \left(\frac{600}{600 + f_y}\right) d, \qquad f_y \ in \ (MPa).$$

$$x_{max} = 0.75 \, x_b = 0.75 \left(\frac{600}{600 + f_y}\right) d.$$

$$a_{max} = \beta_1 x_{max} = 0.75 \beta_1 \left(\frac{600}{600 + f_y}\right) d.$$

Case (1): $a_{max} > h_f$ (N.A. lies in the web):

$$A_{c(max)} = bh_f + (a_{max} - h_f)b_w$$

$$C_{max} = T_{max}$$

$$0.85 f_c' A_{c(max)} = f_y A_{s(max)}$$

$$\Rightarrow \quad A_{s(max)} = \frac{0.85 f_c'}{f_y} A_{c(max)}$$

$$= \frac{0.85 f_c'}{f_y}\left[bh_f + (a_{max} - h_f)b_w\right]$$

$$\Rightarrow \quad \boxed{A_{s(max)} = \frac{0.85 f_c'}{f_y}\left[bh_f + \left\{0.75\beta_1\left(\frac{600}{600+f_y}\right)d - h_f\right\}b_w\right]}$$

Case (2): $a_{max} < h_f$ (N.A. lies in the flange):

$$A_{c(max)} = ba_{max}$$

$$C_{max} = T_{max}$$

$$0.85 f_c' A_{c\,max} = f_y A_{s(max)}$$

$$\Rightarrow \quad A_{s(max)} = \frac{0.85 f_c'}{f_y} A_{c(max)} = \frac{0.85 f_c'}{f_y} b\,a_{max}$$

$$\Rightarrow \quad A_{s(max)} = \frac{0.85 f_c'\, b(0.75)\beta_1\left(\frac{600}{600+f_y}\right)d}{f_y}$$

$$\Rightarrow \quad \boxed{A_{s(max)} = \frac{0.6375 f_c'}{f_y}\beta_1\, bd\left(\frac{600}{600+f_y}\right)}$$

Note: Tables are available for $A_{s(max)}$ in terms of f_c' and f_y (but in psi units).

Example 4 :

Design a T beam for the floor system shown in the figure for which b_w and d are given, with $M_D = 270$ kN·m , $M_L = 460$ kN·m , $f'_c = 20$ MPa , $f_y = 350$ MPa , and simple span $= 5.5$ m .

Solution :

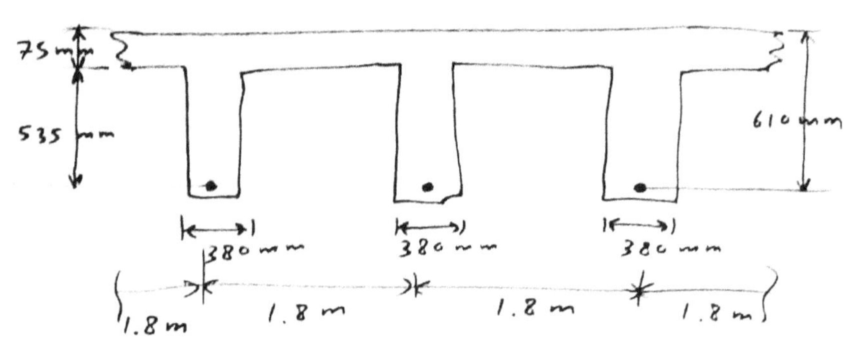

Effective Flange Width :

(1) $\frac{1}{4} L = \frac{1}{4} (5.5) = 1.375$ m $= 1375$ mm .

(2) $380 + (2)(8)(75) = 1580$ mm .

(3) 1.8 m $= 1800$ mm .

$\Longrightarrow$ Take $b = 1375$ mm .

$M_u = 1.4 M_D + 1.7 M_L = 1.4(270) + 1.7(460) = 1160$ kN·m .

$M_u \leq \phi M_n \Longrightarrow M_n \geq \dfrac{M_u}{\phi} = \dfrac{1160}{0.90} = 1289$ kN·m .

Assume $z =$ larger of $0.9d$ or $d - \frac{1}{2} h_f$:

(1) $0.9d = 0.9(610) = 549$ mm .

(2) $d - \frac{1}{2} h_f = 610 - \frac{1}{2}(75) = 572.5$ mm ≈ 573 mm .

$\Longrightarrow$ take $z = 573$ mm (lever arm).

Trial Steel Area :

$$M_n = Tz = f_y A_s z$$

$$1289 = 350^{\times 1000} A_s \left(\frac{573}{1000}\right) \Longrightarrow A_s = 0.006427 \text{ m}^2 = 6427 \text{ mm}^2 .$$

$$C = T$$

$$0.85 f_c' A_c = f_y A_s$$

$$0.85 (20) A_c = (350)(6427) \implies A_c = 132,321 \text{ mm}^2$$

$$\text{Area of flange} = 75 \times 1375 = 103,125 \text{ mm}^2$$

$$A_c > \text{Area of flange}.$$

N.A. lies in the web.

$$A_c = b h_f + (a - h_f) b_w$$

$$132,321 = (1375)(75) + (a - 75)(380)$$

$$\implies a = 152 \text{ mm} > 75 \text{ mm}.$$

Calculate the centroid of A_c :

$$\bar{y} = \frac{(103,125)\left(\frac{75}{2}\right) + (132,321 - 103,125)\left[75 + \left(\frac{152-75}{2}\right)\right]}{132,321}$$

$$= 54.3 \text{ mm}.$$

$$\therefore z = d - \bar{y} = 610 - 54.3 = 556 \text{ mm}.$$

Re-calculate A_s with revised z :

$$M_n = T z = f_y A_s z$$

$$1289 = (350 \times 1000) A_s \left(\frac{556}{1000}\right)$$

$$\implies A_s = 0.006627 \text{ m}^2 = 6627 \text{ mm}^2.$$

$$C = T$$

$$0.85 f_c' A_c = f_y A_s$$

$$0.85 (20) A_c = (350)(6627)$$

$$\implies A_c = 136,439 \text{ mm}^2 > \text{Area of flange}.$$

N.A. lies in the web.

$$A_c = b h_f + (a - h_f) b_w$$

$$136,439 = (1375)(75) + (a - 75)(380)$$

$$\Rightarrow \quad a = 163 \text{ mm} > 75 \text{ mm}.$$

Calculate the centroid of A_c:

$$\bar{y} = \frac{(103,125)\left(\frac{75}{2}\right) + (136,439 - 103,125)\left[75 + \left(\frac{163-75}{2}\right)\right]}{136,439}$$

$$= 57.4 \text{ mm}.$$

$$\therefore \quad z = d - \bar{y} = 610 - 57.4 = 553 \text{ mm} \quad \left(\text{very close to } 556 \text{ mm}\right).$$

Re-calculate A_s with revised z:

$$M_n = Tz = f_y A_s z$$

$$1289 = (350 \times 1000) A_s \left(\frac{553}{1000}\right)$$

$$\Rightarrow \quad A_s = 0.006665 \text{ m}^2 = 6665 \text{ mm}^2.$$

Check ACI Code requirements: $A_s \leq A_{s(max)}$:

since N.A. lies in the web, we use the formula.

$$A_{s(max)} = \frac{0.85 f_c'}{f_y}\left[b h_f + \left\{0.75 \beta_1 \left(\frac{600}{600 + f_y}\right) d - h_f\right\} b_w\right]$$

$$= \frac{0.85(20)}{350}\left[(1375)(75) + \left\{0.75(0.85)\left(\frac{600}{600+350}\right)(610) - 75\right\}(380)\right]$$

$$= 8158 \text{ mm}^2.$$

But $A_s = 6665 \text{ mm}^2 < 8158 \text{ mm}^2$.

Also check $\rho_{min} \leq \rho_w$;

$$\rho_w = \frac{A_s}{b_w d} = \frac{6665}{380 \times 610} = 0.0288 = 2.88\%.$$

$$\rho_{min} = \frac{1.4}{f_y} = \frac{1.4}{350} = 0.004 = 0.4\% < \rho.$$

$\therefore$ Section satisfies ACI Code requirements.

(4) <u>Design of T Beams for Negative Moments:</u>

* When T beams resist negative moments, their flanges will be in tension and the bottom of their webs will be in compression.

* In this case, the rectangular beam design formulas will be used.

* Requirements of ACI Code / Section 10.6.6 :

  <u>Part of the flexural steel in the top of the beam in the negative-moment region should be distributed over the effective width of the flange, or over a width equal to one-tenth of the beam span, whichever is smaller.</u>

* Should the effective width be greater than one-tenth of the span length, the ACI Code requires that some additional longitudinal steel be placed in the outer portions of the flange. The intention of this part of the Code is to minimize the sizes of the flexural cracks that will occur in the top surface of the flange.

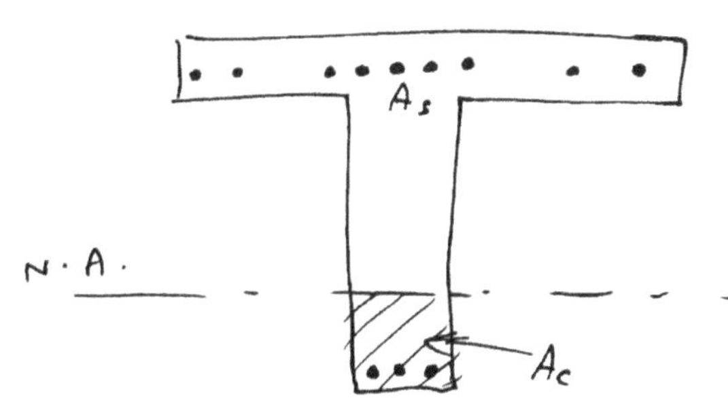

(5) <u>L-Shaped Beams :</u>

* Assume that <u>L-beams</u> (T-beams with a flange on one side only) are not free to bend laterally.

* Handle them in this chapter as symmetrical sections.

* For L-beams, the effective width of the overhanging flange may not be larger than one-twelfth the span of the beam, 6 times the slab thickness, nor

one half the clear distance to the next web. $\boxed{\text{IV}}$
(ACI Code/ Section 8.10.3) .

Design Session V

Serviceability — Deflections

① <u>Introduction</u>:

* Structural design profession is concerned with a <u>limit states</u> philosophy.

* The term <u>limit state</u> is used to describe a condition at which a structure or some part of a structure ceases to perform its intended function.

* There are two categories of limit states: strength and serviceability.

* <u>Strength limit states</u> are based on the safety or load-carrying capacity of structures and include buckling, fracture, fatigue, overturning, and so on.

* <u>Serviceability limit states</u> refer to the performance of structures under normal service loads and are concerned with the uses and/or occupancy of structures including deflections, cracking, and vibrations. These items may disrupt the use of structures but do not involve collapse.

② <u>Importance of Deflections</u>:

* Deflection problems are more important for slender members.

* Excessive deflections for concrete members may cause sagging floors, excessive vibrations, and even interference with the proper operation of supported machinery.

* Deflections may damage a structure's appearance or frighten the occupants of the building, even though the building may be perfectly safe.

③ <u>Control of Deflections:</u>

* Deflections of reinforced concrete members are usually controlled by limiting the member thicknesses in some proportion to their span lengths or by providing maximum permissible computed deflections for various situations.

<u>Minimum Thicknesses</u> (for Beams and one-way Slabs):

ACI Code, Table (9.5(a))

$L \equiv$ Span.

Member	Minimum thickness, h			
	Simply Supported	One end Continuous	Both ends Continuous	Cantilever
	Members not supporting or attached to partitions or other construction likely to be damaged by large deflection			
Solid One-way Slabs	L/20	L/24	L/28	L/10
Beams or Ribbed One-way Slabs	L/16	L/18.5	L/21	L/8

<u>Maximum Deflections</u>:

* If the designer chooses <u>not</u> to meet the minimum thicknesses given in ACI Table 9.5a, he or she must compute deflections.

* If this is done, the values determined may not exceed the values specified in ACI Table 9.5(b).

<u>Camber</u>:

* The deflection of reinforced concrete members may also be controlled by "cambering".

upward camber

MAXIMUM PERMISSIBLE COMPUTED DEFLECTIONS

Table 9.5(b)

Type of Member	Deflection to be Considered	Deflection Limitation
Flat roofs not supporting or attached to nonstructural elements likely to be damaged by large deflection.	Immediate deflection due to live load	$\dfrac{L}{180}$
Floors not supporting or attached to nonstructural elements likely to be damaged by large deflection.	Immediate deflection due to live load	$\dfrac{L}{360}$
Roof or floor construction supporting or attached to nonstructural elements likely to be damaged by large deflections	that part of the total deflection after attachment of nonstructural elements (sum of the long-time deflection due to all sustained loads and the immediate deflection due to any additional live load	$\dfrac{L}{480}$
Roof or floor construction supporting or attached to nonstructural elements not likely to be damaged by large deflection		$\dfrac{L}{240}$

④ Calculation of Deflections:

* the midspan deflection of a uniformly loaded simple beam is <u>five times</u> as large as the midspan deflection of the same beam if its ends are fixed.
* Nearly all concrete slabs are continuous, and their deflections fall somewhere in between the above two extremes.

* For most practical purposes, it is sufficiently accurate to calculate the midspan deflection of a member as though it is simply supported and to subtract from that value the deflection caused by the average of the negative moments at the member ends.

Examples of Deflection Formulas:

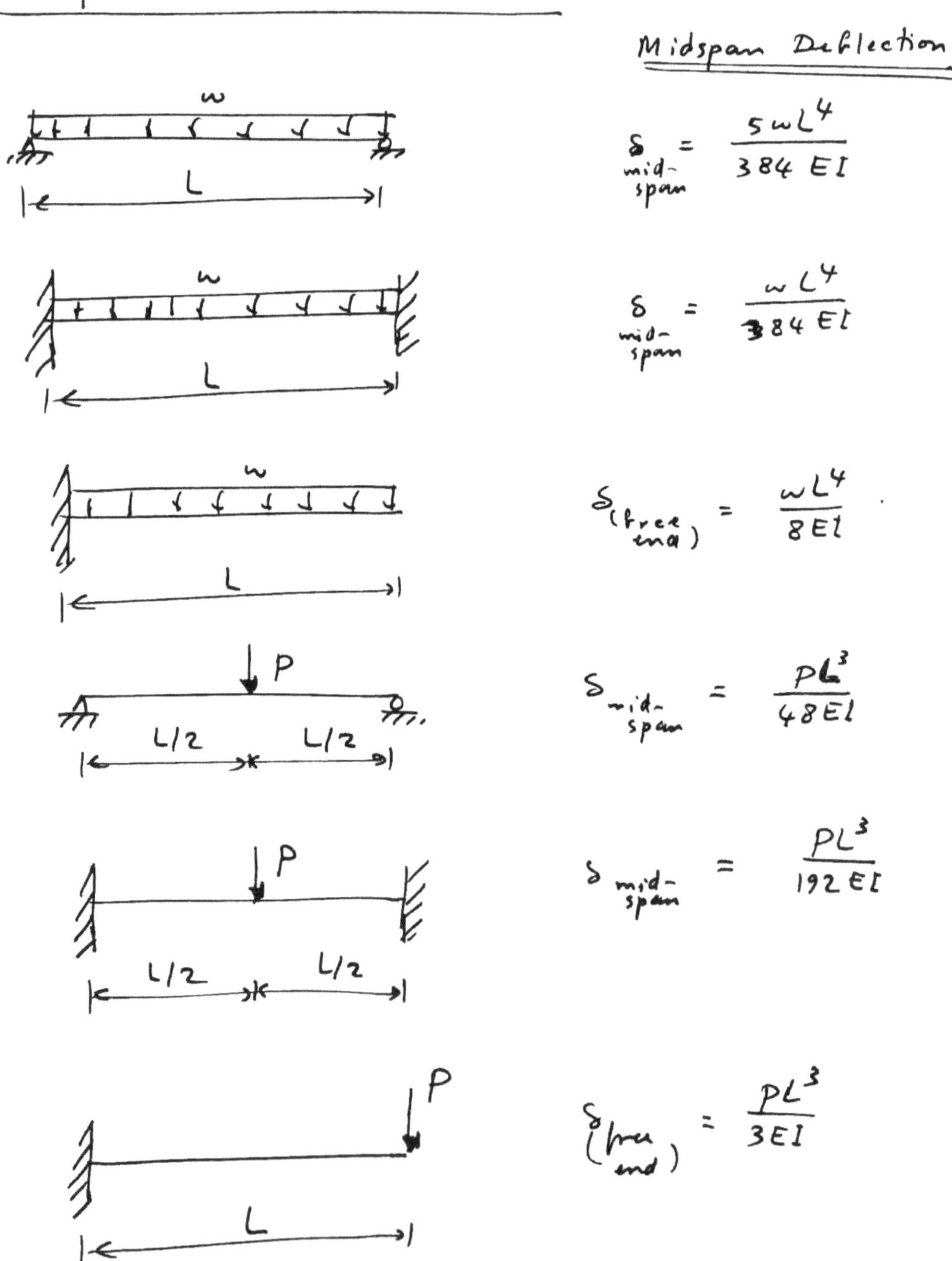

⑤ Effective Moments of Inertia:

* The amount of cracking that has occurred affects the value of the moment of inertia.

* When $M < M_{cr}$, no tension cracks occur, and the full uncracked section provides rigidity, and the moment of inertia I_g is available.

* Although a reinforced concrete beam may be of constant size (prismatic) throughout its length, for deflection calculations it will behave as though it is composed of segments of different sizes (nonprismatic).

* For the uncracked section, we use the gross moment of inertia, I_g.

* For the cracked section, we use the transformed value, I_{cr}.

* For an uncracked prismatic member, use I_g.

* For a cracked beam, we cannot use I_g or I_{cr}. We need a more exact value of I.

* Use the effective moment of inertia I_e.

$$I_e = \left(\frac{M_{cr}}{M_a}\right)^3 (I_g) + \left[1 - \left(\frac{M_{cr}}{M_a}\right)^3\right](I_{cr})$$

ACI Equation 9-7

* The above formula for the effective value, I_e, is given in the ACI Code / Section 9.5.2.3 .

* The above formula is an average value.

* $M_a \equiv$ maximum service-load moment .
  $M_{cr} \equiv$ cracking moment, from $f_r = \frac{M_{cr} \, y}{I_g}$.

  I_g : gross moment of inertia (neglecting steel).
  I_{cr} : transformed moment of inertia of the cracked section.

---

(6) <u>Long-Term Deflection</u> :

* With I_e and appropriate deflection expressions, instantaneous or <u>immediate deflections</u> are obtained.

* <u>Long-term</u> or <u>sustained</u> loads cause large increases in these deflections due to <u>shrinkage</u> and creep.

* The factors affecting deflection increases include :
  - humidity
  - temperature
  - curing conditions
  - compression steel content
  - ratio of stress to strength
  - age of the concrete at the time of loading .

* If concrete is loaded at an early age, its long-term deflections will be greatly increased.

* Excessive deflections in reinforced concrete structures can very often be traced to the early application of loads.

* Because of the several factors mentioned above, the magnitudes of long term deflections can only be estimated.

* ACI Code / Section 9.5.2.5 states that to estimate the increase in deflection due to these causes, the part of the instantaneous deflection that is due to sustained loads may be multiplied by the empirically derived factor λ and the result added to the instantaneous deflection.

$$\lambda = \frac{\xi}{1 + 50\rho'}$$

(ACI Equation 9-10).

* In the above expression which is applicable to both normal and lightweight concrete, ξ is a time dependent factor that may be determined from the following table (ACI Code / Section 9.5.2.5):

Duration of Sustained Load	Time dependent factor ξ
5 years or more	2.0
12 months	1.4
6 months	1.2
3 months	1.0

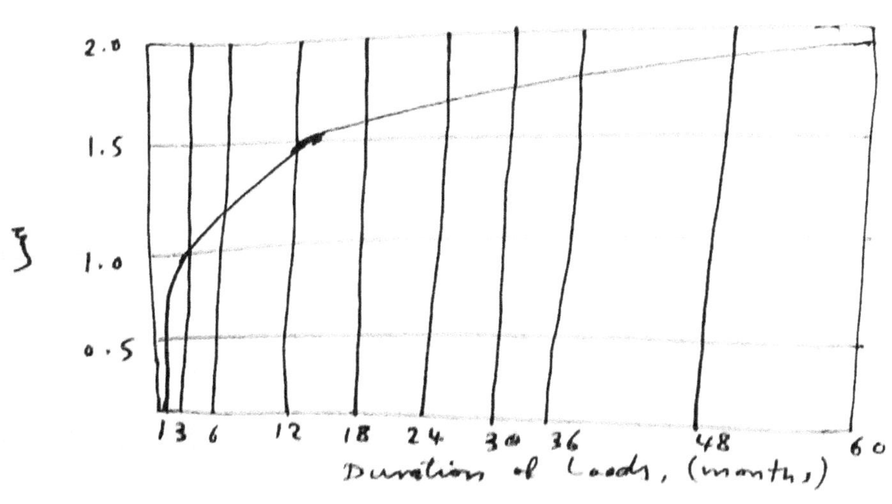

* The effect of compression steel is taken into account in the λ expression with the term ρ', where

$$\rho' = \frac{As'}{bd} \qquad \text{(compression steel ratio)}.$$

* ρ' is to be computed at midspan for simple and continuous beams and at the supports for cantilevers.

---

* The full dead load of a structure can be classified as a sustained load, but the type of occupancy will determine the percentage of live load that can be called sustained.

* For an apartment house or an office building, perhaps only 20% to 25% of the service live load should be considered as being sustained, whereas perhaps 70% to 80% of the service live load of a warehouse might fall into this category.

---

⑦ Simple-Beam Deflections:

Example 1:

The beam shown in the figure has a simple span of 6 m and supports a dead load including its own weight of 14.5 kN/m and a live load of 10.2 kN/m. Use $f_c' = 20$ MPa.

Calculate the long-term deflection.

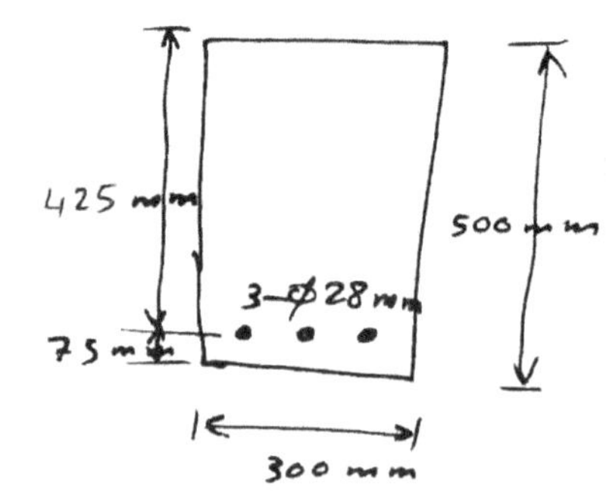

Solution:

(a) Instantaneous Deflection:

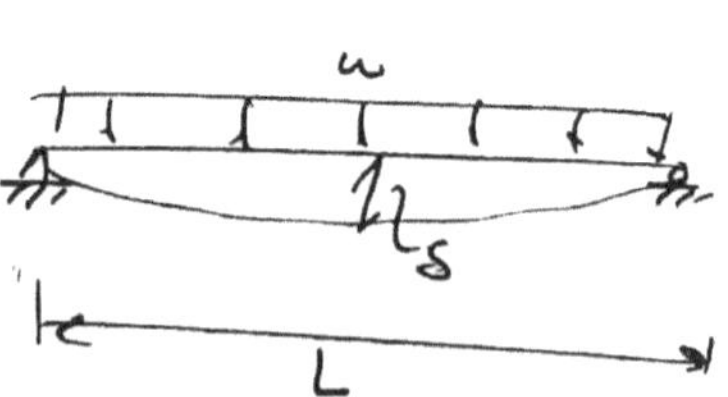

$$I_g = \frac{1}{12}(300)(500)^3 = 3125 \times 10^6 \text{ mm}^4.$$

$$f_r = 0.62\sqrt{f_c'} = 0.62\sqrt{20} = 2.77 \text{ MPa}. \qquad \delta = \frac{5wL^4}{384E_c I_e}$$

$$f_r = \frac{M_{cr}\, y}{I_g} \implies 2.77 \times 1000 = \frac{M_{cr}\left(\frac{0.5}{2}\right)}{3125 \times 10^6 \times 10^{-12}} \qquad \boxed{\text{IV } 9}$$

$$\implies M_{cr} = 34.6 \ kN \cdot m.$$

$$w = w_D + w_L = 14.5 + 10.2 = 24.7 \ kN/m \quad (\text{service load}).$$

$$M_a = \tfrac{1}{8} w L^2 = \tfrac{1}{8}(24.7)(6)^2 = 111.15 \ kN \cdot m.$$

Using the transformed area method:

$$A_s = 3\left[\frac{\pi (28)^2}{4}\right] = 1847 \ mm^2.$$

$$A_t = n A_s$$

First calculate n :

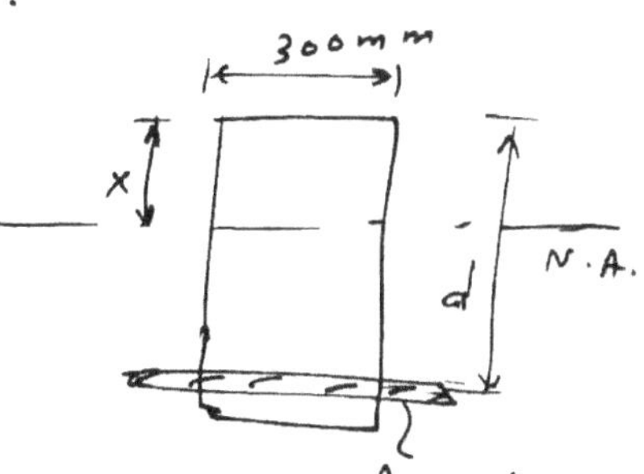

$$E_c = 5000 \sqrt{f_c'} = 5000\sqrt{20} = 22,361 \ MPa.$$

$$n = \frac{E_s}{E_c} = \frac{200,000}{22,361} = 8.94 \longrightarrow 9$$

$$\text{Take} \quad n = 9 \qquad (\text{or use the table})$$
$$(\text{see Chapter 4}).$$

$$\therefore \ A_t = 9(1847) = 16623 \ mm^2.$$

Take moment of areas about the N.A. :

$$300 x \left(\frac{x}{2}\right) = 16623\,(425 - x)$$

$$150 x^2 + 16623 x - 16623(425) = 0$$

$$x = \frac{-16623 \pm \sqrt{(16623)^2 - 4(150)(-16623)(425)}}{2(150)}$$

$$= 169 \ mm.$$

For the cracked section:

$$I_{cr} = \tfrac{1}{3}(300)(169)^3 + (16623)(425 - 169)^2$$

$$= 1572 \times 10^6 \ mm^4.$$

$$I_e = \left(\frac{M_{cr}}{M_a}\right)^3 I_g + \left[1 - \left(\frac{M_{cr}}{M_a}\right)^3\right] I_{cr}$$

$$= \left(\frac{34.6}{111.15}\right)^3 \times 3125 \times 10^6 + \left[1 - \left(\frac{34.6}{111.15}\right)^3\right] \times 1572 \times 10^6$$

$$= 1619 \times 10^6 \ mm^4 .$$

$$\delta_{inst.} = \frac{5 w L^4}{384 E_c I_e} = \frac{5(24.7)(6)^4}{(384)(22364 \times 1000)(1619 \times 10^6 \times 10^{-12})}$$

$$= 0.0115 \ m$$

$$= \underline{\underline{11.5 \ mm}} .$$

(b) **Long-term Deflection:**

$*$ Calculate the instantaneous deflection due to sustained loads.

Sustained Loads $\nearrow$ 100% DL $\Rightarrow$ 14.5 kN/m

$\searrow$ 30% LL $\Rightarrow$ 0.3(10.2) = 3.06 kN/m.

Sustained Load = 14.5 + 3.06 = 17.56 kN/m.

instantaneous deflection due to sustained load = $\frac{17.56}{24.7}(11.5)$

$$\therefore \quad \delta_{sust.} = 8.18 \ mm .$$

For five years or more, $\xi = 2.0$

$\rho' = 0$ (no compression steel):

$$\lambda = \frac{\xi}{1 + 50\rho'} = \frac{2.0}{1 + 50(0)} = 2.0 .$$

Long-term deflection $= \delta_{inst.} + \lambda \delta_{sust.}$

$$= 11.5 + (2.0)(8.18)$$

$$= 27.86 \ mm$$

$$\approx \underline{\underline{28 \ mm}} .$$

⑧ Continuous—Beam Deflections:

* Consider a continuous T beam subject to both positive and negative moments.

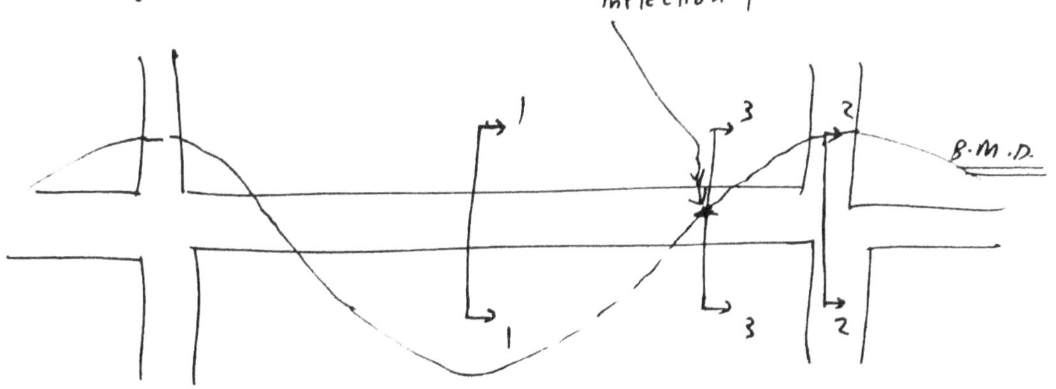

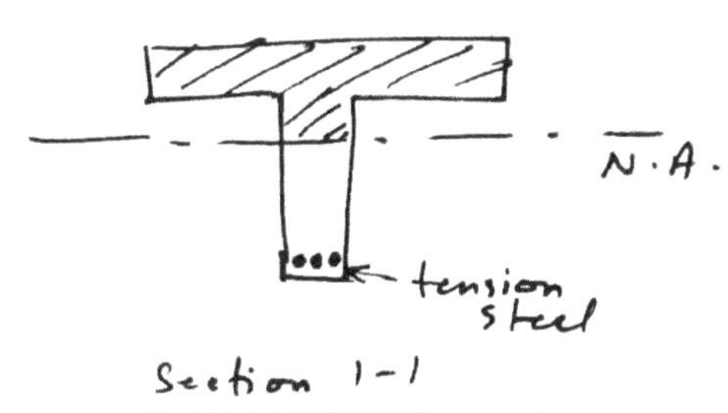

Section 1-1
+ve moment

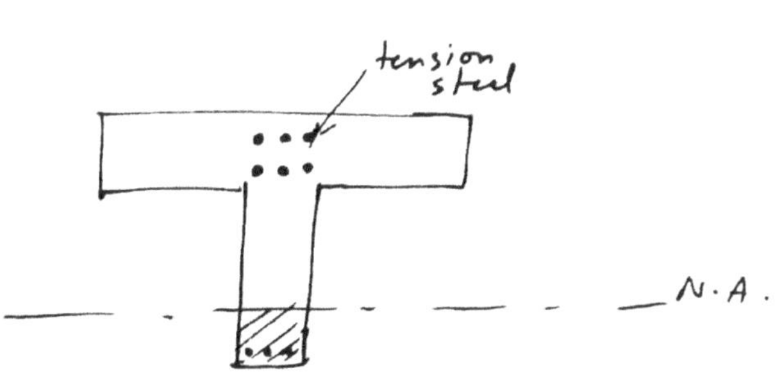

Section 2-2
—ve moment

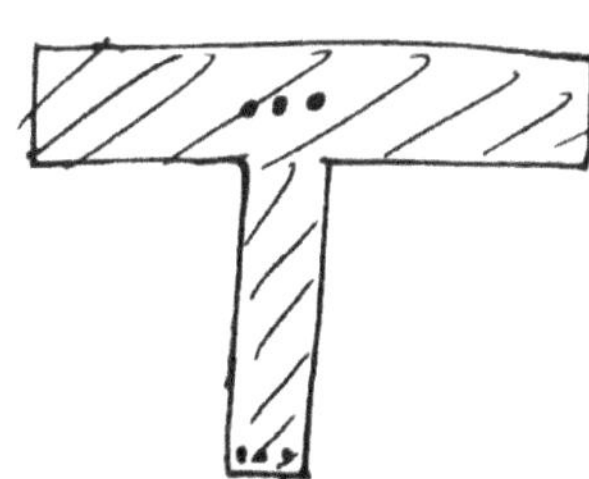

Section 3-3
Near Inflection Point
($\approx$ zero moment)
usually $M < M_{cr}$.
(no cracks in
this section).

* ACI Code / Section 9.5.2.4 permits the use of a constant moment of inertia throughout the member equal to the average of the I_e values computed at the critical positive- and negative- moment sections.

* the multipliers at the two sections should be averaged for long-term calculations.

Example 2:

Determine the instantaneous and 5-year deflections at the midspan of the continuous T beam shown in the figure. The member supports a dead load including its own weight of 21.9 kN/m and a live load of 36.4 kN/m, of which 50% is assumed to be sustained. Use $f_c' = 20$ MPa and $n = 9$. The moment diagram for the full dead and live loads is shown in the figure along with the beam cross-section.

Solution:

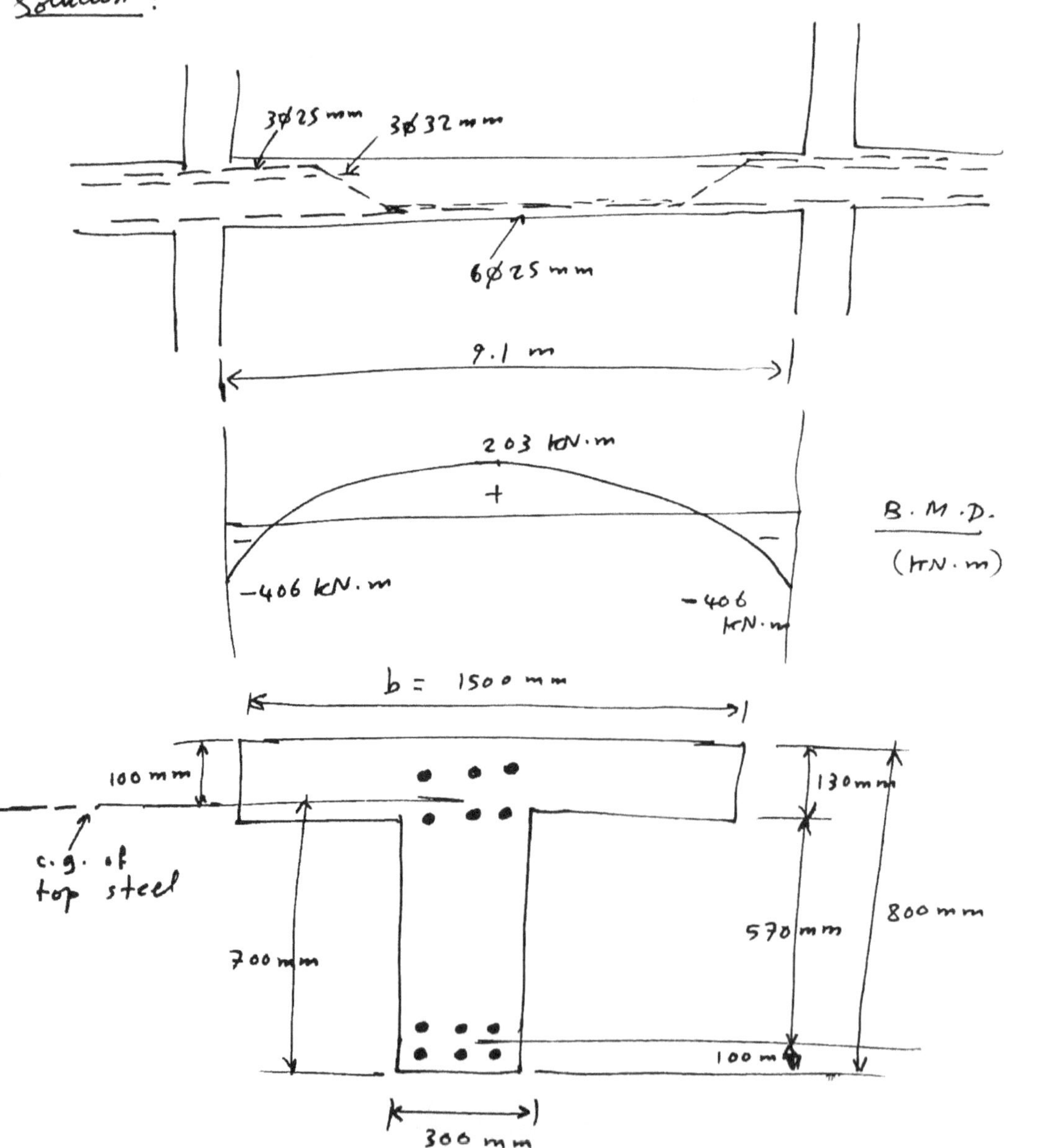

1 <u>For Positive – Moment Region:</u>

$$\bar{y} = \frac{(130)(1500)\left(\frac{130}{2}\right) + (300)(670)\left(130+\frac{670}{2}\right)}{(130)(1500) + (300)(670)}$$

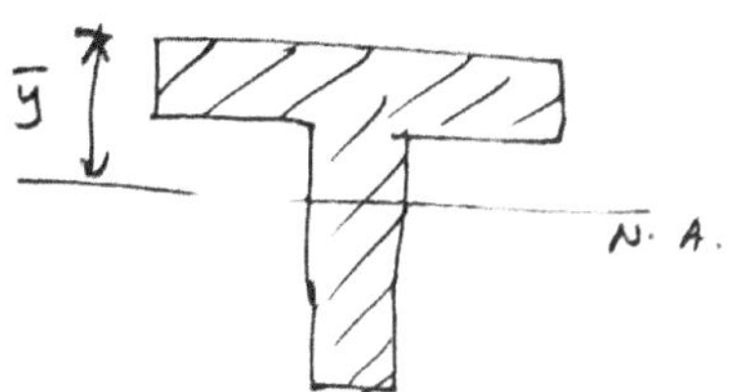

$$= 268 \text{ mm}.$$

$$I_g = \frac{1}{12}(1500)(130)^3 + (1500)(130)\left(268-\frac{130}{2}\right)^2$$

$$\quad + \frac{1}{12}(300)(670)^3 + (300)(670)\left(800-268-\frac{670}{2}\right)^2$$

$$= 23{,}630 \times 10^6 \text{ mm}^4.$$

$$f_r = 0.62\sqrt{f_c'} = 0.62\sqrt{20} = 2.77 \text{ MPa}.$$

$$f_r = \frac{M_{cr}\,y}{I_g} \implies 2.77\times1000 = \frac{M_{cr}\times(800-268)/1000}{23630\times10^6 \times 10^{-12}}$$

$$\implies M_{cr} = 123 \text{ kN}\cdot\text{m}.$$

Consider the transformed section:

$$A_s = 6\left[\frac{\pi(25)^2}{4}\right] = 2945 \text{ mm}^4.$$

$$A_t = nA_s = 9(2945) = 26{,}507 \text{ mm}^4.$$

Take moments of area about N.A.

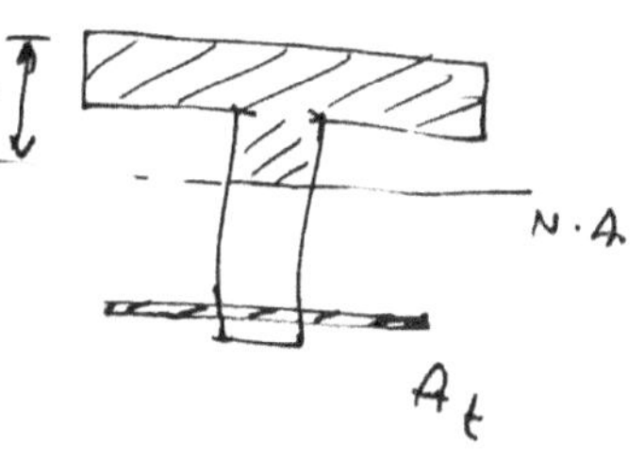

$$(1500)(130)\left(x-\frac{130}{2}\right)$$

$$+ (x-130)(300)\left(\frac{x-130}{2}\right)$$

$$= 26507(700-x)$$

$$(1500)(130)x - (1500)\frac{(130)^2}{2} + 300(x-130)\left(\frac{x+130}{2}\right) = 26507(700-x)$$

$$(1500)(130)x - 1500\frac{(130)^2}{2} + 150(x^2-(130)^2) = 26507(700-x)$$

$$150x^2 + 221507x - 33764900 = 0$$

$$\therefore \quad x = \frac{-221507 \pm \sqrt{(221507)^2 - 4(150)(-33764906)}}{2(150)}$$

$$\therefore \quad x = 139 \text{ mm} \quad > 130 \text{ mm.} \qquad \checkmark \quad \underline{\underline{o.k.}}$$

$$I_{cr} = \frac{1}{12}(1500)(130)^3 + (1500)(130)\left(139 - \frac{130}{2}\right)^2$$

$$\qquad + \frac{1}{3}(300)/139 - 130)^3 + 26,507\,(700 - 139)^2$$

$$\qquad = 9685 \times 10^6 \text{ mm}^4.$$

$$M_A = 203 \text{ kN·m}.$$

$$I_e = \left(\frac{M_{cr}}{M_a}\right)^3 I_g + \left[1 - \left(\frac{M_{cr}}{M_a}\right)^3\right] I_{cr}$$

$$\qquad = \left(\frac{123}{203}\right)^3 \times 23630 \times 10^6 + \left[1 - \left(\frac{123}{203}\right)^3\right] \times 9685 \times 10^6$$

$$\qquad = 12,787 \times 10^6 \text{ mm}^4.$$

$\boxed{2}$ $\quad \underline{\text{For Negative-Moment Region:}}$

For the gross-section, consider a rectangular section.

$$\bar{y} = \frac{800}{2} = 400 \text{ mm.}$$

$$I_g = \frac{1}{12}(300)(800)^3$$

$$\qquad = 12,800 \times 10^6 \text{ mm}^4.$$

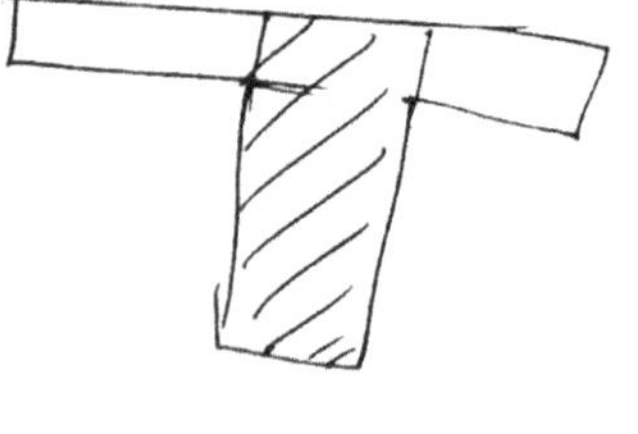

$$f_r = \frac{M_{cr}\,y}{I_g} \implies 2.77 \times 1000 = \frac{M_{cr}\,(400/1000)}{12800 \times 10^6 \times 10^{-12}}$$

$$\implies M_{cr} = 88.64 \text{ kN·m}$$

Consider the transformed section:

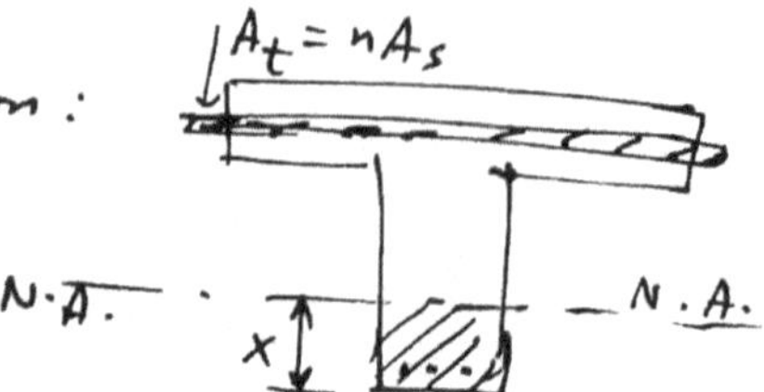

$$A_s = 3\left[\frac{\pi(23)^2}{4}\right] + 3\left[\frac{\pi(32)^2}{4}\right]$$

$$\qquad = 2885 \text{ mm}^2.$$

$$A_t = n A_s = 9(2885) = 34968 \text{ mm}^4.$$

Take moments of area about N.A.:

$$300 \times \left(\frac{x}{2}\right) = 34968 \,(700 - x)$$

$$150 x^2 + 34968 x - 34968(700) = 0$$

$$x = \frac{-34968 \pm \sqrt{(34968)^2 - 4(150)(-34968)(700)}}{2(150)}$$

$$\therefore x = 304 \text{ mm}.$$

$$I_{cr} = \frac{1}{3}(300)(304)^3 + 34968(700 - 304)^2$$

$$= 8293 \times 10^6 \text{ mm}^4.$$

$$M_a = 406 \text{ kN} \cdot \text{m}.$$

$$I_e = \left(\frac{M_{cr}}{M_a}\right)^3 I_g + \left[1 - \left(\frac{M_{cr}}{M_a}\right)^3\right] I_{cr}$$

$$= \left(\frac{88.64}{406}\right)^3 \times 12\,800 \times 10^6 + \left[1 - \left(\frac{88.64}{406}\right)^3\right] \times 8293 \times 10^6$$

$$= 8426 \times 10^6 \text{ mm}^4.$$

---

3 <u>Instantaneous Deflection</u> :

$$\text{Average } I_e = \frac{1}{2}\left[\left(\frac{8426 \times 10^6 + 8426 \times 10^6}{2}\right) + 12787 \times 10^6\right]$$

$$= 10,606.5 \times 10^6 \text{ mm}^4.$$

$$E_c = 5000 \sqrt{f_c'} = 5000 \sqrt{20} = 22,361 \text{ MPa}.$$

To calculate the deflection at midspan, use the <u>moment-area theorems</u> (or any other method):

δ = moment of $\frac{M}{EI}$ diagram between C and B about B

$= \frac{2}{3}\frac{(609)}{EI}\left(\frac{9.1}{2}\right) \cdot \frac{5}{8}\left(\frac{9.1}{2}\right)$

$- \left(\frac{406}{EI}\right)\left(\frac{9.1}{2}\right)\left(\frac{9.1}{4}\right)$

$= + \frac{1051}{EI_e}$

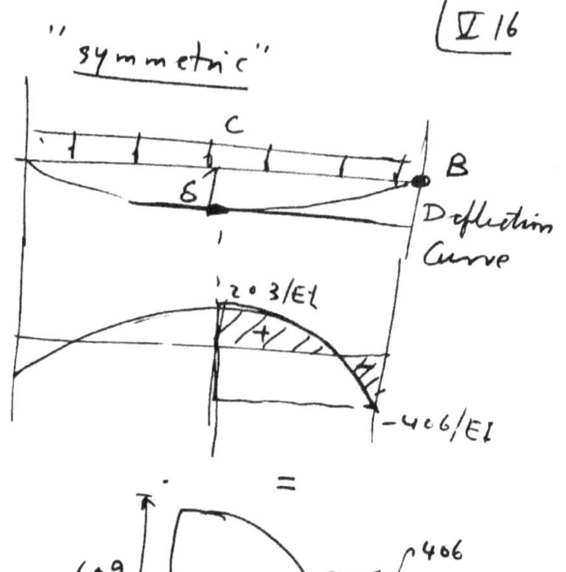

$\therefore \quad \delta_{\substack{mid-\\span\\(inst.)}} = \dfrac{1051}{(22361 \times 1000)(10666.5 \times 10^{6} \times 10^{-12})} = 0.0044\ m$

$= 4.4\ mm.$

$\boxed{4}$ **Long-Term Deflection:**

Sustained Load $= 1.0\ DL + 0.5\ LL$

$\qquad = 1.0\ (21.9) + 0.5\ (36.4)$

$\qquad = 40.1 \quad kN/m.$

$\delta_{sust.} = \dfrac{40.1}{21.9 + 36.4}\ (4.4) = 3.03\ mm.$

For positive-moment region, $\rho' = 0$:

$\lambda = \dfrac{\xi}{1 + 50\rho'} = \dfrac{2.0}{1 + 50(0)} = 2.0$

For negative-moment region, $\rho' = \dfrac{A_s'}{bd} = \dfrac{3\left[\frac{\pi(25)^2}{4}\right]}{(300)(700)}$

$\qquad\qquad = 0.00701.$

$\lambda = \dfrac{2.0}{1 + 50\,(0.00701)} = 1.48.$

Average $\lambda = \frac{1}{2}\left[\left(\frac{1.48 + 1.48}{2}\right) + 2.0\right] = 1.74$.

⌊Ⅴ⌋7

Long-term deflection $= \delta_{inst.} + \lambda\,\delta_{sust.}$

$$= 4.4 + (1.74)(3.03)$$

$$= 9.67 \text{ mm}.$$

<u>Note</u> :

* the ACI Commentary ($R\ 9.5.2.4$) says that for approximate deflection calculations for continuous prismatic members, it is satisfactory to use the midspan section properties because these properties which include the effect of cracking have the greatest effect on deflections